SAFETY AUDITING
A Management Tool

SAFETY AUDITING
A Management Tool

Donald W. Kase
Kay J. Wiese

VNR VAN NOSTRAND REINHOLD
_____ New York

Copyright © 1990 by Van Nostrand Reinhold

Library of Congress Catalog Card Number 90-47381

ISBN 0-442-23746-4

All rights reserved. No part of this work covered by the copyright heron may be reproduced or used in any form or by any means—graphic, electronic, or mechanical, including photocopying, recording, taping, or information storage and retrieval systems—without written permission of the publisher.

Manufactured in the United States of America

Published by Van Nostrand Reinhold
115 Fifth Avenue
New York, New York 10003

Chapman and Hall
2-6 Boundary Row
London, SE1 8HN

Thomas Nelson Australia
102 Dodds Street
South Melbourne 3205
Victoria, Australia

Nelson Canada
1120 Birchmount Road
Scarborough, Ontario M1K 5G4, Canada

16 15 14 13 12 11 10 9 8 7 6 5 4 3 2 1

Library of Congress Cataloging-in-Publication Data

Kase, Donald W.
    Safety auditing : a management tool / by Donald W. Kase, Kay J. Wiese.
    p. cm.
    Includes index.
    ISBN 0-442-23746-4
    1. Industrial accidents—Investigation. 2. Industrial safety—Auditing. I. Kase, Donald W. II. Title.
HD7262.25.W54 1990
658.4′08—dc20

90-47381
CIP

Contents

Preface	ix
Introduction	xiii
Part I Management Expectations	**1**
1 Systems and Styles	3
Safety—A Management Tool?	3
Safety Auditing—A Management Tool	5
Responsibility, Authority, and Accountability	7
Proactive, Reactive, and Passive Management	11
The Purposes of Auditing—Management Expectations	18
Influencing Management	22
2 Theory Versus Practice	24
Recognizing Philosophies and Styles	24
Dealing with Various Philosophies	27
Interpreting Actions and Decisions	29
Large/Small Contrasts	31
Client and Employer Relationships	32
3 Successful Auditing	36
Operational Success	36
The Successful Manager	39
The Successful Safety Auditor	40
Safety Auditing—A Management Training Tool	41
Successful Communication with Management	43
4 Involving Line Management	46
Educating Management	46
Interviewing the Operating Manager	49
Tailoring the Interview	55
Coaching and Improving	59

Part II Planning and Preparation — 63

5 Knowing the Hazards — 65
Knowing the Operations — 65
Comprehensive and Specific Audits — 68
Reviewing the Paperwork — 69

6 Prioritizing — 73
What Can Go Wrong? — 73
The Maximum Credible Event (MCE) — 77
What Can Cause an MCE? — 80
Likelihood and Seriousness — 82
Targeting the Right Hazards — 85

7 Knowing the Requirements — 90
Hazards, Opinions, and Requirements — 90
Legitimizing Requirements — 93
The General Duty Clause — 96

8 Standards — 100
State and Federal OSHA — 100
In-House Standards — 106
Advisory Standards and Codes — 107
Contractual Standards — 109
Deriving a Checklist — 110

Part III Effective Safety Audits — 113

9 Protocols and Practices — 115
Announced or Unannounced? — 115
Introductions and Intentions — 118
Findings—Positive and Negative — 121
Documentation — 123
Eliciting Solutions — 126
Reinforcement — 128

10 Evaluating Management — 131
Measuring Involvement and Commitment — 132
Management Activities — 135
Employee Perceptions — 143
Management Systems — 151
Delegation — 152
Discussion of a Management Evaluation Checklist — 154
How To Use the Checklist — 161

11 Evaluating Equipment and Facilities	167
First Impressions	167
Checklists	170
Generic Checklists	197
Field Notes	202
Citing Requirements	204
Recommendations	205
12 Evaluating Work Practices	207
Procedures	207
Compliance	211
Field Notes	215
Management Acceptance	223
Improvements	225
Recommendations	229
13 Communications	232
Real Time	233
Explaining the Problem	235
Acknowledgements	238
The Final Report	240
Other Audits	242
Part IV Analysis	**245**
14 Evaluating Management Performance	247
Twisting the Tiger's Tail	249
Credibility	251
Recommendations	256
15 Identifying Physical Deficiencies	261
Documentation	261
Endemic or Epidemic?	265
The Real Findings	269
Management Accountability and Actions	271
16 Identifying Work Practice Deficiencies	275
What Are They?	276
Behavioral Observations	278
Endemic or Epidemic, Part 2	281
Management Accountability and Actions	285
17 Trend Analysis	289
Deficiencies and Accidents	290
Record Keeping and Retrieval	293

Spotting Trends	299
Analytical Approaches	301
Raising the Red Flag	305
18 Epilogue	308
Index	311

Preface

As we approach the 21st century, the need for proactive (the opposite of reactive) loss prevention is greater than ever before.

We have produced an explosion of new technology involving highly toxic products in complex new operations. Man's ventures into space, new information processing systems, myriad products for defense, and a glittering array of new consumer items are only a few of the new technological areas that challenge our best loss prevention efforts.

Industrial safety and health has also been affected by catastrophic events. The Sunshine Mine—Bhopal—Chernobyl—are events that sensitized the United States (and indeed the world) to the necessity for effective loss control to protect people and property. Public pressures and outrage stemming from these catastrophies have reinforced the necessity for proactive loss control.

In addition to new technology and social catastrophe, a "world economy" has impacted industrial safety and health. Intensified industrial competition has narrowed profit margins in the United States. These narrowed margins have made it less and less acceptable to lose profit dollars due to accidents. In addition, the dollars spent for safety programs are under more scrutiny, and the effective use of these dollars is critical.

In 1970, a new law dramatically affected industrial safety and health. The American worker was given the right of a "safe workplace" as a matter of national policy. The Occupational Safety and Health Act established standards together with civil penalties for noncompliance. In the latter 1980s, banner headlines emphasized OSHA's enforcement actions: "COLLAPSE OF HIGH RISE

BRINGS 5.11 MILLION FINE FROM OSHA"—"I.B.P. INC. GETS 2.6 MILLION OSHA FINE." These serious fines further reinforce the need for effective and proactive controls to ensure compliance with the law.

The year 1985 set very serious precedents for industrial safety and health. In a landmark case involving an Illinois company, management personnel were charged with murder after the death of a worker from on-the-job cyanide poisoning. The President, Plant Manager, and Foreman were found guilty of industrial homicide. Largely as a result of this case, prosecutors in various parts of the United States now investigate workplace injuries and/or fatalities for potential criminal liability. This potential criminal liability adds to the need for effective loss prevention.

Besides potential civil and criminal liabilities, the need for proactive loss control has been accelerated by far-reaching medical discoveries and knowledge about occupational illness during the last forty years. As an example, products like asbestos and PCB's have become household words. Their potential for damaging human health is well known. During this same period, the work force has become better educated and more assertive around issues involving worker safety. These two factors have combined to accelerate the need for proactive control of workplace hazards.

In addition, the last forty years have seen the growth and then decline of labor unions. Even though union strength has diminished in recent years, unions are still a formidable force in areas of industrial safety and health. In companies with unionized labor forces,, the union adversarial strength adds to the need for proactive industrial safety and health.

Skyrocketing litigation with massive settlements are adding to the need for effective loss control methods. In many states, injured employees and their attorneys are finding avenues for suing companies in spite of Workmen's Compensation Laws. Also, third-party suits have become commonplace.

Finally, accelerating insurance rates add to the need for effective loss control. Indeed, both insurance rates and insurance availability are causes for concern as we approach the 21st century. All of these—new technology, Bhopal, the world economy, civil/criminal liability, litigation, and insurance costs—provide compelling reasons for proactive loss control.

Most would agree that we live in an exciting, mind-boggling age. At the same time, it is an age with massive challenges. More than ever before, companies must have effective loss control that utilizes proactive methods. It is not enough to cope with accidents after the fact. For all of the reasons listed earlier, we must *act* to comply with laws, *act* to protect people and property, and *act* to preserve the profitability of our companies. We believe that *safety auditing* is an important and proactive tool for loss control. This book is about that tool!

Our book has been written over the past year and a half, and it, like many nonfiction efforts in our modern world, might very well be nearly obsolete by the time it is read. Nevertheless, it is offered as a snapshot of how we have come to understand and conduct safety audits in our recent professional endeavors. There are numerous facets and scopes of industrial safety auditing, and neither of us will ever claim to have thought of all of them.

If this work serves to stimulate some ideas for building safety into our readers' operations, new or old, then it has been a gratifying experience. We make no claim, however, that it will ever prepare the safety professional, nor operating management, to discover and/or solve every potential safety problem that might exist.

Acknowledgments

This book is dedicated to the people who have fostered and encouraged our professional growth. Without them, we undoubtedly would have languished in the pits of mediocrity, and never have had the gumption to speak out against the things we disdain, nor to publicly applaud and promote the things we believe in. Among the numerous individuals we would wish to acknowledge, the following are at the forefront of our thoughts:

A. A. Imus	F. A. DeMaestri
J. W. White	H. F. McQueen
P. F. Preston	W. L. Childress
J. McDermott	H. B. Chare
T. F. Davidson	T. T. Pinder

These individuals have influenced and shaped our professional lives in a way that no one, other than we, can comprehend. Our thankfulness and appreciation for them cannot be expressed in mere words.

Don Kase
Kay Wiese

Introduction

Loss control is indeed a growing concern and goal in Corporate America, and throughout the industrialized world. It has been estimated that as much as fifteen percent of our Gross National Product in the United States is lost to accidents, disabilities, and injuries. This is a terrible waste of property and human capabilities, not to mention the pain, suffering, and loss of life that accompanies these losses.

Safety auditing, as we will develop it in this work, is a structured and detailed approach to reducing and controlling the seriousness of accidents, hopefully before they occur. It is not a novel idea to prevent accidents. It is not a novel idea to conduct safety inspections. What is different and fairly new about safety auditing is the depth to which accident prevention and containment efforts, and floor inspections, are carried, and the probing into detail that some seemingly superficial issues can instigate. It is a way to evaluate management, the work force, *and* the physical plant itself. It offers a disciplined and structured approach to plant inspections and evaluations, and more.

Another major difference between audits and traditional inspections is the time element. A safety inspection might be measured in terms of hours, while we talk about safety audits of an operating area or department in terms of weeks. The difference is in the operative word "thorough."

This book is written to provide a basic understanding of the philosophy, politics, methods, protocols, and use of the data derived from safety audits, as we have developed and evolved it in our work experience. It is intended for the safety professional and enlightened operating manager alike—for those who wish to upgrade the inherent safety (that is, reduce losses) in their operations.

We write from an industrial background. Our point of view is that of the safety professional in a modern-day industrial enterprise. We do not presuppose that those in a similar situation are the only people who might pick up this book, so we hope that the principles, philosophies, and approaches we describe will be generic enough to be applied in any environment, if not the nitty-gritty methodologies.

We view successful safety auditing as consisting of four major facets, or elements, and only one of those is auditing per se. First, and in our opinion, the foundation of the whole undertaking, is the development of an empathetic understanding of who operating management is, how they think, how different managers behave and relate to others, what drives them and what their priorities are. Part 1 of this book deals entirely with management styles, practices, and expectations. Over the years, we have encountered all stereotypes, and most intermediate shades, from the tyrannical straw boss to the timid pleader, from the commando-style take-charge manager to the oblivious or arrogant politico. We have learned that it is of crucial importance to the safety auditing effort that auditors recognize what type of management they are dealing with, and how they function. After all, it is the managers who will or will not pay attention to the audit findings, who will or will not implement corrections and improvements. For maximum effectiveness, the interface between auditors and managers should be tailored to the managers' traits, not to the auditors' particular wishes.

The second element, detailed in Part 2 of this book, is the thorough preparation, by the safety auditors, of themselves. Presuming to conduct plausible safety audits, and generating positive responses to safety problems, yet knowing next to nothing about the operations being audited, will literally kill the auditors' credibility and effectiveness. We espouse spending the time necessary, conceivably as much time as the audit itself, but surely a substantial period, becoming thoroughly familiar with the operations, their hazards, the materials and equipment involved, the potential consequences of plausible accidents, and what forces might drive workers toward unsafe choices.

We try to maintain some kind of distinction between legalistic requirements and safe operating practices and safe equipment and facilities. Part of the business of adequately preparing for an area audit is researching the applicable regulations, both internal and external, as well as sound advisory standards and consensuses. This aspect, knowing what is required, *and* knowing what is best, is stressed in Part 2.

Part 3 is the how-to portion of the work. To do a thorough audit, there are three aspects that need to be evaluated. Safety audits, just like the traditional safety inspections, assess the physical appearance and condition of the operating area. We would never belittle this facet of a safety audit, and a major segment of Part 3 is devoted to this aspect. However, as we said, it is only one aspect of the total safety posture. It only reflects one-third of the cultural whole.

Some critics may complain that this book will be a vendetta on industrial management, because we consistently saddle them with the responsibility for safety problems and their correction. It is not a vendetta, but we do frequently observe that they are the only ones who can make a difference. Another major portion of Part 3, Effective Audits, deals with the evaluation of management systems, policies, and requirements, and their implementation in the real world of the plant. Their involvement, communicativeness, openness and accessibility, and their willingness to improve and achieve excellence are auditable traits.

The third aspect of actual auditing is the assessment of work practices on the floor. The neatest package of management policies and dictums, combined with an impeccably conceived and maintained plant, can be totally defeated by sloppy and noncompliant operational practices. We will try to illustrate how to separate noncompliant operational practices from those that cannot be complied with.

Finally, we suggest, in Part 4 that myriads of audit findings are virtually worthless unless understood in the context of the big picture. The summarization of countless individual findings, both positive and negative, into a meaningful package that a proactive management can comprehend and deal with, is at least as important as those findings themselves. A single isolated safety problem that is corrected and then forgotten may, indeed provide a measurable, localized benefit to the safety of that particular operation. However, unless generic problems are dealt with in a generic fashion, and true excellence is observed, encouraged (even applauded), and rewarded, the corrective measures taken will be treating only the symptoms of the real problems.

We want this book to be a thought stimulator, not a handbook of do's and don'ts. We acknowledge up front that the ideas, methods espoused, purposes, and values of this relatively new discipline are ours alone. We do not wish to implicate anyone as having endorsed or supported the premises, concepts, methods, or conclusions that we have put to print. There are those in the profession whom we admire and have learned from, and we thank them for their help in straightening out our own thoughts.

Therefore, safety auditing is presented herein as a four-part effort —identifying and dealing with management types, knowing and understanding the operations to be audited, conducting the audit itself on the management, physical plant, and work practice levels, and, finally, what to do with the audit findings to effect the maximum positive response.

PART I

Management Expectations

If there is one thing that management *wants* from safety auditors, it is help, not harassment and trouble. Whether they *expect* help or harassment is not immediately controllable by safety auditors, but whether they *get* help, and not a pain in the neck, is controllable by, and the responsibility of, only the safety auditors.

This part of the book is placed at the beginning because we believe that the entire process of planning, preparation, execution and reporting, and the analysis of safety audit findings, should be developed, learned, and understood from the perspective of providing help to operating management, enhancing the success of the enterprise, and (in a spirit of self-preservation) avoiding the possibility of being perceived as "trouble."

Safety auditors can make positive and helpful contributions by controlling the methods and manner by which they do their job. First of all, their product—their very reason for being safety auditors—is information. They provide information to the management that they serve, just as the quality control, accounting, and process engineering people do. Information is a neutral commodity. It is neither holy nor evil. It can seem to become evil if it is presented in a derogatory or accusative manner. It can be polished up to seem laudatory when no plaudits are deserved. The first lesson that safety auditors should master is that they should never report anything but neutral information.

The second service they should render to operating management is in identifying strengths of the operation, upon which management can capitalize. One fatal error for safety auditors is to ignore those islands of excellence that do in fact abound in this world. One comment on an observed good work practice, or a neat and orderly operating area, or a foreman who makes an extra effort to try to create a team, will provide a model and a foundation upon which to replicate such strengths.

Conversely, to be sure, safety auditors should identify weaknesses and deficiencies that need improvement. Many think of this as the only function of safety auditing, a thought that we are trying very hard to dispel. However, it is necessary to present the neutral facts of unsafe conditions or practices. Safety auditors should try to relate the strengths that they find to the possible solutions to the weaknesses and deficiencies. If one area of excellence can be expanded to solve a deficiency, it will be much easier to accomplish, than if some whole new thought process or discipline has to be developed.

Another way in which safety auditors can help management solve safety problems is by stimulating teamwork. In this context, we envision the "team" as being composed of all interested parties, including the foremen and workers, management, and the safety auditors. Whenever they can possibly promote ownership of the situation, they can expect that the most creative thought will go into finding the best solution. Safety auditors might have to be the unofficial sponsors of the team approach, but that's okay. They can create a team while no one even knows there is one. They just elicit problem-solving thinking from everyone with whom they interface.

A motto we like to refer to often, and challenge some of our critics with occasionally, is, "strive for excellence, not just compliance." Then we add, with a wink, "compliance is a good start, though."

In Part 1, we will discuss management's various, and sometimes bewildering, ways of managing, how they tend to view themselves, safety auditors and safety as a whole. We will speculate on what roles various kinds of managers see for themselves in safety. We will talk about how safety auditors can learn to determine, early on, what type and style of manager they are dealing with. Some thoughts will be offered on the effect of the size of the operating organization, and whether the manager is a fellow employee or a client. We will look into the concept of operational success, successful management, and successful safety auditing. Finally, the vital elements of management involvement, and communication with them, will be explored.

Chapter 1

Systems and Styles

The primary instinct of safety auditors is to observe "what is," and report it against "what should be." This chapter is intended to acquaint novice safety professionals who may never have considered such mundane and subprofessional questions before, with the real world of operational management.

One manager will receive a critical, probably negative, finding, with utter joy—that a potentially disastrous problem has been illuminated, and can be immediately corrected. More likely than not, another manager will receive that same potentially catastrophic message with a mixed assortment of indignation, skepticism, defensiveness, or aggressive counterattack.

It is vital that novice safety auditors learn to understand the roles and limitations of both operating management and themselves. The nearly infinite variety of operating managers demands an equally infinite variety of interface practices by safety auditors.

In this chapter, we will discuss the purpose and expected results of safety auditing when confronting the various styles of operating managers that might be encountered. To do so, we will first describe those various stereotypes that might be lurking out there for unwary auditors.

Finally, we will talk about how to communicate with, and communicate in an influential manner with, the several types of managers that might be seen on the operating floor.

Safety—A Management Tool?

Yes, safety is a management tool. Like all tools, some are rusty, some are sharp, some are never used, and some are used to their fullest. Some are very simple, some are quite complex, and some don't work at all.

There are some who tend to think of safety as a program that is heaped on top of a job to be done, rather than as a part of the job. We describe and plan the job, then maybe figure out whether and how we can do it safely. This separation into two thought processes will inevitably result in tacked-on safety, rather than built-in safety. When safety is an add-on, it is not a management tool. If a carpenter goes out to build a house with saws, hammers, nails, lumber, tape measures, and safety (hard toes, safety glasses, gloves, helmet, etc.), that person will likely produce a quality house at a reasonable cost and not get hurt. If the carpenter tries to build without the saw, the house will probably not get built, or it will be built at a staggering cost. If all of the measurements are eyeballed, rather than made using a tape measure, a house may be produced, but the quality will certainly suffer. Finally, if safety is left behind, the chances are the house won't get built either, because of the cost to the carpenter's own well-being.

Whenever we have done something successfully, it has been because safety was an integral part of the process of doing. It was one of the tools that dovetailed with all of the other tools, materials, machines, and people that created that success. This is true even when we are oblivious to it, when it is not a conscious act on our part to "be safe." The fact is, accidently or on purpose, safety is present in the accomplishment, or it would not have been accomplished.

This is not to say that there was not an element of luck, or that there were not hazardous conditions or actions that we managed to blunder through safely, or even that maybe we took a chance at Russian roulette and won. But if the job was successful, it was done safely.

We could say the same thing for quality, skills, and talents too. If the lumber is rotten, or the "carpenter" is a preapprentice novice, the venture is not likely to be an overwhelming success, either. If the job was done successfully, it was done by people who knew how to do it, with materials that were up to specification. So we can expand our formula for success to include the right combination of talents, materials, and tools safely integrated to do the job.

We use simple tools like a knife and fork, a pencil, or a door knob virtually without thinking. When was the last time you had to consciously think your way through the intricacies of cutting paper with scissors? Other tools are more complicated. The effort we put into using tools is a direct function of their complexity, as well as the frequency of use. If safety is to be an effective, useful, and valued management tool, then our efforts, both as managers and safety professionals, have to be tuned to the complexity and hazards of our operations, and our safety activities built to match them.

Safety is indeed an essential component of success. You can never enjoy success without having somehow been safe. The more success is jeopardized by hazards and operational complexity, the more sophisticated the safety tool must be, and the more we have to use it.

It is from this perspective — that safety must be one of the tools of a successful operation — that safety professionals and operating management can begin to formulate their approach to safety as a whole, and to safety auditing and analysis. It will help define both planning and execution, and help safety auditors, or safety engineers who are planning to conduct audits, to understand management's expectations. If it is viewed and discussed as a tool, rather than a program aside from operations, it will begin to become integrated into the operations. Safety is a production tool; safety auditing is the balance sheet.

Safety Auditing — A Management Tool

While safety may be fairly described as a proactive management tool, and we just said that safety auditing is the balance sheet — the measure of the effectiveness of the safety disciplines — we propose that safety auditing is also an extremely proactive management tool, when it is instituted by the top management of an enterprise. Executives who charter safety auditing functions in their organizations have declared to everyone in those organizations that they mean business when it comes to safe production.

On the other hand, safety professionals, who are proposing to top management that safety auditing functions be established, have a selling job to do. Selling safety auditing as a management tool will take some thought and preparation. Safety professionals must make their case for safety auditing in competition with the production manager, who wants some new equipment, the finance director, who has to have some new computers, and the sales manager, who cannot survive without more salesmen and lower prices for the products.

Presumably, the company has a safety program and safety engineers in place in the organizational structure. The CEO is not going to automatically understand and accept the notion that now a safety auditing staff ought to be funded, with the probable travel expenses that that entails, and the nearly inevitable fact that safety auditors will find problems that will cost money to correct. The value of safety auditing to management is a direct function of how committed that management is to minimizing their losses through accidents and fines. It is a given that a safety auditing program will add direct costs, and it is virtually impossible to account for the savings that will accrue from it.

The best (and only) sales pitch safety professionals have at their disposal is the one that insurance salesmen use. If you want to minimize your losses, you had better know what the situation is out in the real world, and act on that information. Ignorance is neither an excuse nor much solace after the worst happens.

Safety auditing is a method whereby management can receive a continuing evaluation of its safety effectiveness. It can be structured so as to be dynamic in its feedback to management. It is not a snapshot, but rather a movie, of the safety discipline in the organization. Repeated safety audits of an operation begin to draw a picture of the safety-consciousness of that operating group. The findings from one audit to the next can be a very revealing contrast. The expectation is that findings will be on a continuous decline, as operating management learns from past mistakes. The degree to which such learning has occurred will be measured in subsequent audits. This has to be valuable information to the top levels of management.

Safety auditing is, at the same time, a sampling procedure. It is patently impossible for safety auditors to obtain a 100 percent review of any operation. They cannot be on the floor 24 hours a day, all week, and for months at a time. So there is an element of statistical evaluation involved. Operators may be on their best behavior when safety auditors are in the vicinity. So be it. When auditors report that a certain operational behavior or discipline was observed, and laud it, it will become much more fashionable for both operators and their immediate management.

Safety auditing is a tool by which enlightened management seeks to know the truth. If they are, indeed, interested in knowing the truth, then the facts reported by safety auditors will be received as heads-up information, upon which positive action can be taken. If the management is not interested in knowing the truth of their safety situation, then safety auditing will serve little or no purpose.

This book is intended to prepare safety professionals for initiating or directing safety auditing programs. It is also aimed at operating managers who truly want to cut their losses by integrating safety into their operations. We will get into much detail about how to evaluate management systems and practices. We will really get down to the nitty-gritty in inspecting physical plant operations, and comparing "what is" with "what should be." We will look into the problematical, even mystical, aspect of worker's practices vis-à-vis what they are trained to do, or what their operating procedures tell them to do. We do not purport to have all the answers to making an operation accident- and injury-free. We do suggest that the auditing protocols and procedures we will develop here will illuminate the situation as it exists, and offer methods for improvement.

Responsibility, Authority, and Accountability

Let's suppose that you have a seat at ringside for a heavyweight bout where the world champion is about to defend his title. Both fighters enter the ring, the announcements are made, and the bell sounds. Each combatant moves aggressively toward the center of the ring, when you see that one fighter is shackled with handcuffs. Would you expect the handcuffed fighter to win?

We've all seen situations basically like the one above in industrial safety. But first, let's take a look at three, very important, words: responsibility, authority, and accountability.

How can one be *accountable* for production when he has been given no *responsibility*? What sense does it make to hold someone responsible for a condition, action or thing over which he has no *authority*?

These three words (responsibility, authority, and accountability) are of major importance to safety professionals. We need to know the meaning and relationship between these words to understand management styles. In addition, these three words are vital when we are evaluating the environment for safety auditing. We also need to know what management means when they use these words. In ordinary conversation, and even in management policies, the words "responsibility, authority, and accountability" are often used vaguely and even interchangeably.

The dictionaries tend to run us around in a circle too. Responsibility is defined as "the state, quality or fact of being responsible." It can also be "duty, obligation or burden." Authority is given as "the power to command, enforce laws, exact obedience, determine or judge." (American Heritage Dictionary, 2nd College Edition, Houghton-Mifflin Co., Boston MA) There appears to be an inherent and discernible difference between duties and power. Accountable is defined as "answerable," so accountability is the situation where one has to answer for something. But that brings us back to our duties or obligations.

We have all seen and read policy statements in which some manager is made responsible for a well-defined end result, but the policy ends there without any statement of his "authority" to command the necessary resources.

Let's look at these three words (responsibility, accountability, and authority) in the context of building a house. Suppose you are responsible for building the house we talked about earlier. You are responsible for completing it by a given date and the finished house must match the blueprints. In addition, you must meet the previously determined cost and have all of the systems work. Water must come from the hot water faucet. The kitchen light must turn on by the kitchen door, not from the bathroom, etc.

Suppose further, that you are denied the authority to select your starting date. Besides that, you are denied the right to specify the materials and supplies to be used; and you may not choose where to buy those materials. In addition, you are barred from choosing the crew or even deciding what size crew is needed for the job. As a final roadblock, you may not determine what tools will be used.

The decision-making authority is retained by the buyer, and he wants summer jobs for all of his kids. This means that construction can't start until school is out in June. Besides that, the buyer dictates that materials for the house must come from a second-hand supply store owned by his cousin. Under these conditions, can you hope to fulfill your responsibility?

Obviously not. Where someone is given responsibility without authority, it is virtually impossible for him to succeed and be effective. The house-building project fails. Now who is accountable for its failure? You hold the buyer responsible (or is it accountable?), for imposing impediments to your performance. He holds you accountable for failing to meet the schedule, quality, and cost conditions.

Now what?

Implied in accountable, or answerable, is the idea that the accounting or answer is demanded by a higher authority. In the building example above, the authority is probably a building inspector and/or a judge. In industrial safety, it may be a federal agency (OSHA), a state Industrial Commission, or a review judge (national, state, or local).

At first glance, the building example cited earlier may seem extreme. Yet, we've all seen instances where effective safety performance is impeded by some very tough roadblocks. An individual is assigned the responsibility for safety in a given area and is held accountable for negative events. An accident, or an OSHA inspection resulting in citations, is viewed as evidence of incompetence. However, the individual's responsibility and accountability are never teamed with authority. The person simply doesn't have the clout to get the necessary resources (people, time, money) to do the job effectively. So let's take a look at these three words (responsibility, accountability, authority) and define them as they operate in an industrial environment, with regard to safety.

Responsibility for safety is the obligation to provide safe conditions, tools, equipment, procedures, practices, and materials to execute the job. So who has that responsibility? Obviously, the responsibility is shared by everyone who contributes to getting the job done. This includes technical input from the safety people, the tool/facility designers, the procedure writers, and even the product designers. It also includes the people who move materials, maintain equipment, and schedule the jobs and crews. Finally, there are the people who actually do the work, make the product, package, and store it, together with all of the foremen and managers.

So everyone is responsible, which too often means that *no one is responsible.* All too often, this is a cop-out written into management policy and the policy is used to excuse lethargy.

So who *is* responsible?

Individuals must be responsible for accomplishing their jobs in a safe manner. If a person's job is running a machine, that person must be properly trained. After training, the person must undertake to operate that machine in a manner that produces quality products safely and efficiently. To sum it up, the machinist must be responsible for that operation. If, on the other hand, a person's job is running a company, or a plant, or a department, that person too must undertake to operate the organization (operation) in a manner that is safe, effective, and efficient—therefore profitable.

In other words, safety responsibility goes hand-in-hand with operating responsibility—they are inseparable. *You are responsible for the safety of whatever you are responsible for—period.*

As we illustrated in the house-building project earlier, responsibility with no authority is a contradiction in logic. It's like putting handcuffs on a boxer and then expecting him to win his bout. When management assigns responsibility, but withholds the authority to get needed resources (people, money, time), the responsibility will likely go unfulfilled. A front line foreman has no authority to execute safety responsibilities, if the foreman has to get approvals from three levels of management above in order to get a machine guarded. Similarly, if the safety engineers' recommendations have to filter through layer upon layer of managerial approvals, their authority is nonexistent as well. *Whoever approves has the authority.*

Sometimes management structures itself so that authority resides only in the upper echelons of that structure. In other, more enlightened organizations, decision making and authority are pushed down to the lowest levels. Under the first structure, management runs the show. In the second organization, management guides and advises those who run the show. It is essential, however, that authority and accountability be matched under either of the above organizations. Those who approve or disapprove must be the ones to account for their approvals and disapprovals.

As stated earlier, accountability infers a higher authority. The person at the top must hold immediate subordinates accountable, and they must do the same with their subordinates, etc. If floor operations authority lies at Level Five, then Level Four must judge the wisdom, prudence, timeliness, effectiveness, and safety of Level Five's decisions and actions. Also, after that evaluation, Level Four needs to cheer on, guide, direct, and advise Level Five to facilitate successful and effective performance.

By the same token, Level Three must judge how effectively Level Four executes their planning, directing, advising, and guiding activities. So responsibility, authority, and accountability progress up the management chain.

So let's recap. Responsibility is a duty, a job to be done with certain conditions imposed. Authority is the power to make it happen. And accountability is the report card that scores how well we did the job!

The question for safety auditors at this point is, simply, what do they do when they encounter a gross mismatch among these three concepts? A lack of responsibility with attendant accountability is untenable. A lack of authority without the necessary accountability is likewise intolerable. Also, surely the fact of accountability without the obviously requisite authority is a no-win situation. Safety auditors can only, as we have stated before, and will again, report what is. Their only position is that "here is a foreman (or area supervisor, or department manager, etc.), who is expected to do "X," but is only authorized, by policy, to do "Y," and this situation defeats safety excellence."

There is little that safety auditors can do, unilaterally, to correct policy inequities, other than report those inequities. It is forever up to management to provide the necessary wherewithal for operating, or floor, management, and to correct the safety problems that safety auditors may identify. If the top management of an enterprise is interested in upgrading its safety program, then it will follow that equalization of authority, responsibility, and accountability will ensue. If there is a predisposition toward the expedient, that is, only doing what will get them by today's crisis, then the safety auditors' role is much more difficult, even bordering on the impossible. They are thrust into the role of being accountable, or perhaps responsible, for the correction of safety discrepancies over which they have no control. Beginning safety auditors must ever be on the watch for this potential entrapment.

Proactive, Reactive, and Passive Management

Proactive is a recently coined word that made its way into the dictionary in 1987. Random House defines proactive as "serving to prepare for, intervene in, or control an expected occurrence or situation; especially a negative or difficult one."

An effective fire department is one example of a group who could be described as proactive. Certainly, they plan and train for an expected negative event, a fire. When an alarm comes in, they immediately act to intervene and control the fire, working to keep losses at the lowest possible minimum. In this situation, however, the owner of the burning building is most likely to be passive or reactive. At least, if the owner has tried to understand and control fire hazards in the establishment, these efforts have been ineffective.

Proactive people have a thorough understanding of their operations and valid information about potential problems. From this base, they will clearly define ideals, goals, and objectives; then plan their activities to achieve them. Proactive people will seek to identify, understand, and anticipate potential negative events in order to prevent, or at least control, these events. They work out methods, evaluate their effectiveness, and seek improvements. They scrap activities which have no recognizable, direct, or measurable influence on reaching their objective.

In the best of all worlds, people are made managers because they have been very proactive in managing their operations. They have increased efficiency, reduced costs, and maintained and improved quality. They have likely been successful in matching people to the jobs, and have established stable, friendly relationships with each of them. In short, *they lead*!

If these managers have studied the hazards as thoroughly as they have studied efficiency improvements; if they have been as quick to mitigate hazards as they have been to install labor-saving devices, then they are our kind of manager. They are proactive in safety as well. They accept the responsibility for the safety of their operations, not just the unit costs and output. They understand their ultimate accountability for accidents and their losses, in the same way that they are accountable for costs in labor, scrap, utility, and other components. They exercise their authority to make safety happen.

Proactive managers have another highly significant characteristic; they recognize that safety systems are essential components of production. Further, they know that safety must be an integral part of every production activity, to be fully effective.

Proactive managers do not happen (pardon the expression) accidently. Before they reach this ideal, they first have to mentally make a commitment to achieve safe operations. Their next step is to identify what *they* have to do *personally* and *actively* to fulfill that commitment. They recognize that they will not achieve safe operations by wishing for them while sitting in their oak-paneled office. They must *act*!

In Table 1.1, we have tabulated a list of suggested activities by which a proactive operating manager can be identified. Both safety professionals (who want to introduce this proactive safety concept) and operating managers (who want to improve their operations) can select and adopt some or all of these ideas. This list is only a beginning. Each business enterprise will have its own industry-specific activities for proactive managers.

Working with proactive managers is a dream-come-true for safety professionals. Their role, to advise, cite requirements, and evaluate conditions/practices, is appreciated and acted upon by these managers.

Unfortunately, in safety matters, reactive managers are far more common. In their worst embodiment, reactive managers are completely oblivious to safety problems until after an accident occurs. Then these managers jump in with both feet, and correct everything in sight that is even remotely related to the accident. These items are corrected on a Priority A basis, then once again these reactive managers lapse into unconsciousness with regard to safety. All of their attention returns to production, schedules, costs, scrap rate, personnel problems, reports, and on and on.

Reactive managers see themselves as safety conscious and supportive of safety, even though they have adopted no active role in the safety systems. They measure their support by how quickly and energetically they respond in a crisis. They view good safety performance on their part as minimizing downtime, perhaps disciplining an offender, and finding low-cost (band-aid?) corrections for the conditions and/or work practices that led to the accident. Reactive managers tend to be proud of curing symptoms with localized corrective actions, but they seldom take lessons learned to other portions of their operations. In addition, they don't pursue problems or accidents to the root causes. As a result, only the symptoms, and never the disease itself, gets cured.

Safety professionals who are working with reactive operating managers, have a tougher job than those helping proactive managers. Reactive managers have a limited understanding of their

Table 1.1. A proactive manager's safety activities.

1. *To Understand and Prioritize Hazards*
   a. Reviews, questions, studies, and learns hazards, their causes, and controls.
   b. Prioritizes hazards that have serious injury potential, or significant damage or downtime potential.
   c. Reviews, critiques, questions, challenges, negotiates, and eventually concurs with tool and facility designs.
   d. Reviews, critiques, questions, challenges, negotiates, and eventually concurs in procedures.

2. *To Plan Safe Production*
   a. Reviews necessary operational controls to produce safely, and assesses adequacy/compliance of existing controls.
   b. Assigns corrective actions, and tracks them to completion, for all safety deficiencies identified.
   c. Reviews status in production planning meetings, along with schedules, material availability, repair and maintenance problems, etc.
   d. Consults with floor supervision and safety engineering, makes conscious decisions whether to operate with open safety actions, then obtains higher-level approval.

3. *To Inspect Area and Operations*
   a. Has a written and inviolable MBWA (Management by Walking Around) schedule.
   b. Inspects entire work area twice a week, once during high activity level, once during slack time.
   c. On a rotating basis, inspects one specific machine, work station, or operation in-depth.
   d. Makes written notes, assigns actions, includes in Items 2b and 2c.
   e. Determines root cause of negative observations, to ensure actions are appropriate.
   f. Reports to superior, along with other production statuses.

4. *To Coach in Safety Disciplines*
   a. Explains what is expected, both physically and behaviorally.
   b. Explains why a negative observation is unacceptable.
   c. Explains what correction is needed.

5. *To Motivate/Commend*
   a. In conjunction with 3c and 3d, makes positive observations and commends those concerned.
   b. Keeps records of individual safe performance practices.
   c. Provides recognition of consistent safe performance.

6. *To Investigate Accidents*
   a. Takes the initiative in determining the cause of events, even when Safety and others assist.
   b. Looks for root causes, distinct from proximate causes.
   c. Ensures that timely corrective actions address both cause levels.
   d. Includes such reports in Items 2b and 2c.

7. *Corrective Measures*
   a. Maintains a living log of corrective actions measures, and the status and effectiveness upon completion.
   b. Weekly, reassesses progress, priorities, previous decisions on open items, with his staff and Safety, per Item 2d.

---

responsibility for safety, and feel little accountability for problems or their corrective measures. Often, these managers shift accountability to the safety professionals, while they charge off to run the operations. In these cases, the safety professionals have an educational challenge. They have to define the proactive safety role for the reactive managers, and gradually lead them to an acceptance of responsibility for safety all the time, and not just after an accident occurs.

Passive managers, a third type of manager, are not only oblivious to safety issues, but they never get involved. They expect their safety engineers to take care of things. They can't be bothered—that's Safety's job. In the era of the 1960s and 1970s, passive managers were probably the most common. These managers have the attitude, "I know we have to be safe. That's why I hired a safety engineer—to make sure that we are safe. Now I don't have to worry about that anymore. I can spend my time worrying about important things!"

For passive managers, when an accident happens, their safety engineers have obviously failed. These managers expect Safety to find out what went wrong, to somehow make everything right, and to do all of this at low, or better still, *no* cost. Obviously, in this situation, the safety professionals have nearly impossible, no-win jobs. They have no line authority to obligate resources (people, money, time) for the correction of a problem. All that they can do is advise; and if the operating managers are deaf to their advice, the safety professionals are virtually through. Should they choose to endure, the safety professionals need to document problems, advice, and management inaction for their own self-preservation.

We have discussed three management philosophies and leadership styles: *proactive, reactive,* and *passive.* Proactive operating managers view safety as an essential factor of production, and they *act* to promote, encourage, improve, and reward safety. Reactive managers tend to ignore safety until a problem arises. At that point, they become very active in correcting the causes (or at least crutching them) so that production can resume. Passive managers retain all authority, but make the safety professionals responsible for safe operations.

Obviously there are endless combinations and shades of these three styles of leadership. Indeed, even predominantly proactive managers will have days when they are reactive or passive. Still, safety professionals need to assess the work environment and their managers' dominant leadership styles so that they (Safety) know what point they are starting from, and what their first steps need to be. Safety auditing for passive managers is probably pointless, unless the managers can be educated to at least the reactive level. Audits are like gold to proactive managers, and a source of satisfaction to the safety professionals who conducted them.

We might illustrate the distinction among these stereotypical managers with a hypothetical scenario. Supposing that a handling accident occurs, in which some reasonably valuable material is damaged beyond reclamation or use. It is scrap! It is learned by the manager that Old Tom, who has been the forklift operator in the warehouse for twelve years, ran his tines through a barrel of solvent, and it ran out all over the floor of the loading dock. The barrel of solvent is lost, and the cleanup and airing out of the area took three hours out of the busy warehouse schedule. It will take overtime to catch up, and the company had to pay demurrage on two delivery trucks that had to wait while the cleanup and ventilation were going on. The warehouse manager was disturbed, to say the least.

How would we expect the proactive, reactive, and passive managers to relate with Tom after this incident? Our suspicion is that most warehouse managers would impose some kind of negative consequences, some disciplinary action, on Tom. A verbal tongue-lashing, two days off without pay, a permanent written reprimand, maybe even a demotion, are possibilities. The manager is *reacting* to the accident—*something* is being done about it. "Corrective action" has been taken.

How would we expect a proactive manager to treat Tom. Well, we haven't given all the antecedent facts about the accident, and they are not pertinent to the illustration. However, you can bet that the proactive manager will ferret them out. This manager will first talk to Tom, saying something like, "let's get our heads together and see if we can figure out why this happened, and see if we can figure a way to keep it from happening again." There will be an inspection of the scene of the accident, to find out if there were extraordinary circumstances for which Tom was unprepared or untrained. This manager will work with Tom in a nonaccusatory and nonpunitive way to make sure that there is a thorough understanding of how and why the barrel got punctured. If it turns out that Tom just had a mental lapse, that he was planning a fishing trip, or fretting about the two-foot putt he missed yesterday, the proactive manager will explore and find out whether a new work habit needs to be developed, such as a walk-around of the forklift before each move. Whatever the outcome and action, the proactive manager will make sure that it is a bargain struck between Tom and the manager, not something imposed on Tom.

If, on the other hand, the accident happened because the barrel had been placed in an unsuspected location, and it was too dark at 4:30 A.M. to see it, the manager will take immediate steps to provide adequate lighting. There may have to be a decision made on the necessity for a designated area for solvent barrels, and training for the stockkeepers on where that is. Also, this manager will ask Tom whether there are any other hazards he can think of that they can correct now—before an accident brings it to their attention.

The passive manager will simply send the safety engineer to the scene to find out what happened, fix it, and report back. This noninvolvement in safety considerations is this manager's trademark, and the manager is only interested in hearing that things are back to normal. If told that Tom had a mental lapse, this manager might reply, "well, those things happen," and go on to other, more important, things. If told that it was too dark in that corner of the warehouse at night, and barrels aren't supposed to be there, this manager's instructions will probably be to not put barrels there any more.

Of these three contrasting styles, it should be patently obvious that the only one who corrected an unsafe condition is the proactive, solution-oriented manager. The others either made a flurry of punitive reaction or, in fact, did nothing at all in the way of true corrective action.

The Purposes of Auditing — Management Expectations

A colleague some years ago used to say, "What you expect is what you get!" The statement may be an oversimplification, but we have all experienced the "expectation" phenomenon. Five people attend a meeting and come back with five different versions of what happened and what was said. These different versions are the product of many factors, but individual expectation is prominent. Stated another way, two college students take the same course. One student expects it to be boring, and a waste of time. The second student expects the course to be stimulating and valuable. The major probability is that each student will find what he expects.

This "expectation" factor also has a significant impact on safety auditing, especially where management expectations are involved. Every manager is unique. We have tried to neatly package them as proactive, reactive and passive, while acknowledging that there are infinite combinations of these three qualities. Each manager is going to expect, or hope for, different things.

Operating managers will have one set of expectations for safety audits. Safety managers may anticipate something quite different. The auditors' own bosses will also have specific requirements. Finally, the CEOs, with their overall perspectives, are likely to have their own unique sets of expectations. In addition, if we allow that these management expectations can be either proactive, reactive, or passive, the potential combinations are numerous.

Let's take a look at major expectation categories, starting with the auditors and what they should expect of themselves. Each safety auditor has one person under their direct control, themselves. They must maintain professional perspective and never react on an emotional level to people or events as personal threats. If they can maintain their professional demeanor and exercise their professional skills, auditors have the best chance of persuading management. They should plan and carry out their auditing with the firm intent that they will be factual and objective, regardless of the problems and deficiencies they discover. Safety auditors must never compromise their professional integrity. They never undertake an audit to "nail them to the wall;" and they never whitewash a real problem. The auditors' expectations for themselves must be that of pragmatic purists. If they can maintain their professional stance and communicate effectively, the auditors' job will be a lot easier, their stress levels will be kept under control, and their management contacts will be more receptive to their work.

Also, to be effective, auditors must identify and understand operating managers' expectations of safety audits. If they tend to be proactive, the managers will expect the audits to be valuable tools for upgrading their safety programs and status. It probably won't occur to them that the audits or auditors might be a threat. They will be glad to have unsafe conditions or practices identified so that they can act to eliminate them. Proactive managers will expect fair and positive recognition for good conditions and safe work disciplines. They have put in a lot of hard work to make their operations safe. Most of all, proactive managers, almost by definition professionals, will expect a professional job from the safety auditors.

Reactive managers will expect the audit to result in more work. They may not be afraid of extra work, but they don't go looking for it. They may be dimly aware of some deficiencies, but they are very busy managing costs, schedules, manpower, and their inevitable problems. They don't have time to work on safety issues. Also, reactive managers will view safety auditors as people to be accommodated, people to react to in a vaguely positive, if superficial, way. Afterward, safety audits and auditors can be pushed aside so that the reactive managers can work on important matters.

Safety auditors must not let the reactive managers' response dissuade them from meeting their own expectations of total professionalism. Auditors are seldom welcomed whether they are from the IRS, corporate accounting, or Safety. Safety professionals must be prepared for all kinds of receptions and expectations. When the introductory greeting is "here comes trouble," or "here comes more work," auditors will quickly recognize that they is dealing with reactive managers.

Passive operating managers may feel that they have no time to greet or talk to auditors. They focus their intellect and effort on "important problems" and feel no direct or personal concern that their operations are to be audited for safety. After all, any safety deficiencies are problems that belong to their safety engineers. Indeed, deficiencies are apt to be seen as a negative measure of the safety engineers' competence. If passive managers have any safety audit expectations, they will likely center them on production issues. They are apt to feel concern that the auditors will slow things down and occupy their peoples' time, thus diverting them from production.

Where possible, safety professionals have to be tenacious with passive managers. If the auditors are on the managers' staffs, they may not be able to push hard enough to get audit meetings. If, however, auditors are from outside the company or associated with a different organization in the company, the auditors can strenuously seek personal meetings. Offers to meet with the managers any hour of any day (24 hours a day and seven days a week), will be hard for such managers to refuse.

In addition to the operating managers, safety auditor may need to consider the expectations of in-house safety engineers. In organizations where safety engineers are in plant work areas on a daily (or regular) basis, and safety auditors are from outside the operating department, another interface and set of expectations will emerge.

The area safety engineers will likely have developed a proprietary interest for work areas within their jurisdiction. Indeed, this "owning" and "caring" about the operations add strength to their performance as safety engineers. However, these qualities may mean that they might be somewhat defensive. Certainly, defensiveness is natural in their situation. Presumably, the safety engineers have done their best to make the operations as safe as they can within the limits of their responsibility, expertise, and influence. Independent, outside safety auditors will be evaluating the effectiveness of the safety engineers to some degree, as well as the organizations themselves. Whatever safety shortcomings the auditors identify will, in some way, reflect on the safety engineers, and not just operating management.

Safety auditors must be sensitive to the safety engineers' potential apprehension. Presumably, they are fellow professionals, kindred spirits that speak the same language and have the same goals and ethics. In that case, safety auditors have a real opportunity to create strong professional bonds and lines of communication with the safety engineers. The auditors need to understand the safety engineers' work environment, including the leadership style of top management. They also need to learn about the safety engineers' efforts and acknowledge their successes. Finally, the auditors *need to listen*, so that they can learn the safety engineers' items of concern for their operating areas. Auditors have a unique opportunity to reinforce the safety engineers' efforts, where these efforts have failed to move management to action. Truly professional safety engineers will welcome this kind of help, not fear it.

Everyone has a boss, including safety auditors, and they must recognize the boss's expectations too. In a service-client relationship, the "boss" will be those clients who hired the auditors. In an employee relationship, and depending on the organization, the bosses may be corporate safety directors, plant managers, presidents, or local safety managers. In other industries, the bosses will be corporate audit directors or human resources managers.

Whatever the job title, the "bosses" may or may not be safety professionals. In either case, the boss can either pose a problem, or become an asset, for safety auditors. It depends on the boss's leadership style (i.e., proactive, reactive, or passive). Obviously, proactive bosses will be of great help. They will probably try to learn, and they will expect and appreciate professional work. Passive bosses can be assets too, in that they won't get involved or interfere. Of course, they won't help either.

It is not our intent to detail every personnel interface that safety auditors will encounter. In addition, we don't presume to instruct the reader in all of the ways available to deal with any particular situation. We are hopeful that this discussion will sensitize potential safety auditors to "expectations" as a factor that can significantly impact the success or failure of safety auditing. We hope that this section has illustrated how expectations can vary, and caution safety auditors to understand these expectations before blundering ahead. In short, *know the politics*, because it does enter into the picture.

Influencing Management

In a sense, safety professionals want to influence management much like management wants to influence its work force. Managers want their employees to do specified things in a specified way at a specified time. They expect certain patterns of behavior, and use their influence in a variety of ways to get it. Direction, punishment and reward, reasoning, and training are the vehicles commonly used to influence employees performance, and we see a wide range of mixtures of those, from one organization to another.

The safety professionals' goal, likewise, is to evolve operating management toward a proactive role in safety. They want managers to do certain things, to develop activities, patterns and thought processes that build safety into their operations. But they are not usually in a position to direct, punish, or reward. Education and reasoning are left as the methods available for safety professionals to influence management. There are a number of ways in which safety professionals do, in fact, influence management, whether to the enhancement or detriment of safety. It is our purpose there to explore these ideas and contrast positive from negative influences.

We can see at least four avenues through which safety professionals can try to get their message through. Obviously, their degree of success, and which, if any, avenues are successful, will depend, to a large degree, on the type of managers they are trying to influence. But success depends equally on how skillfully they use these approaches.

By far the most powerful influence on management priorities, decisions, and actions is the economic bottom line. When operating management is convinced that it is in their own, and/or the company's best interest to do something, they will do it. Consequently, when safety professionals, whether as on-the-floor resident safety engineers, or as "outsider" safety auditors, can show an economic advantage to safety improvements, they will usually win their case. The cost of an accident versus the cost of the correction needs to be credible, and the probability of the accident must be supportable. Data is the best evidence that safety engineers can use to make the cost trade-off. Safety engineers should use other experts to help them determine the costs of corrective measures, to maximize credibility. They should also carefully analyze the possibilities of what kinds of accidents, injuries, or property loss incidents can occur if the situation goes uncorrected. This is to ensure that, in the cost of the accident, the exposure is not overstated—again, establishing credibility.

If safety professionals can influence management to act based on economics alone, other approaches, other avenues of persuasion, can become unnecessary. However, regardless of the method employed, credibility is a must. If it is economics, credible numbers. If it is legal compliance, credible citing and interpretation of the requirement. If it is a professional opinion or conclusion, credible technical analysis and evaluation of the alternatives. A calm, analytical assessment of the deficiency or hazard, and a reasoned and supportable analysis of the needed corrective actions, will be far more influential than dictatorial or emotional approaches.

Another helpful hint for influencing management is to offer solutions, not just to identify problems. Even a careful explanation of the problem, one that is totally convincing that corrective measures are needed, will be less likely to stimulate action if no solution is offered. In many cases, the solution is implicit in the problem, but that is not always so. Safety professionals may not be able to engineer the solution, but they can work with operators on the floor, with design or product engineers, with a wide variety of other knowledgeable people, to come up with an effective solution. Coupled with the identification of the problem, an effective solution will probably sell itself.

Another useful approach, that we hope has not, and never will, become jaded with use, is positive reinforcement—catch them doing it right and cheer them on. This mentality in safety professionals can be just as powerful an influence on management as it is when used on lower-level employees. We all thrive on earned praise, from our subordinates, peers, and superiors alike. Even if the managers, whose decisions we are trying to influence, are our bosses, catching them doing it right and applauding will make it a lot easier for them to do it right next time.

All of this sums up to one thing. The interpersonal relationship between safety professionals and operating managers will make or break the influence the former can exercise on the latter. It must be factual, credible, supportive, solution oriented, and cordial—in a word, professional.

Chapter 2

Theory Versus Practice

The experienced safety professional will acknowledge early-on that there is usually a distinct, often tangible difference between management safety theories and policies, and the nitty-gritty world of operational reality. This is a truism the world over, and a universal indictment of the quick-buck culture that, unfortunately, most of us owe our livings to.

In order to operate with the maximum possible effectiveness, safety auditors must learn to recognize the various kinds of operating management with which he might be dealing. It is also crucial that they be able to interpret different managers' reactions and responses to their input.

We will also, in this chapter, discuss the variations on the safety auditing theme that arise when auditing small operations or companies, and (more important then it may at first sound) look at the nuances of the client/consultant relationship, as distinct from the employer/employee relationship presumed throughout the rest of this book. Theories and practices can be widely divergent, even when the theories, and the policies espoused by management, are virtually coincidental. Ferreting out the differences between theory (policy) and practice is challenging and, at the same time, a caution to the budding safety auditor.

Recognizing Philosophies and Styles

Management philosophies vary as people do. Every manager has one, whether articulated or not. Furthermore, the articulated philosophy may not coincide with the practiced style. Clearly, the practiced one is the one that safety engineers or auditors have to recognize and understand, in order to deal effectively with that operating manager.

Many have tried to categorize management practices. McGregor (1960) described Theory X managers, who are policy, rule, and control-oriented, and Theory Y managers, who are facilitators and cheerleaders. Blake and Mouton (1964) developed a grid and quantified managers numerically. Herzberg (1966) distinguished employee-motivational from employee-satisfaction factors, and described a management philosophy that acknowledged and used both to enhance employee performance. It is not our intent here to repeat or analyze these various management concepts, practices, and philosophies, nor to judge their effectiveness. Rather, we intend to contrast various styles and suggest how the safety professional might best deal with them.

In our more prosaic frame of reference, six stereotypical kinds of management philosophies or practices may be envisioned:

1. The tyrant or dictator.
2. The manipulator or schemer.
3. The fussbudget.
4. The commando.
5. The one-man show or loner.
6. The competent, careful journeyman manager.

The safety professional can learn to deal with each of these, but must learn to recognize the kind of person being dealt with.

Dictators may, but usually will not, be interested in advice and counsel. They are probably very adept at delegating assignments and responsibility, or giving orders, but hold on tightly to any authority. They alone make the decisions. They are probably very thorough in their follow-up to assignments that they have made, and are perhaps quite critical of the results. They want facts, not opinions, not expert assistance. It upsets a dictator when subordinates interface with other managers (the manager's peers or superiors). They are very conscious or rank and hierarchy. They alone will speak for their department, and any of their subordinates who presume to do so will quickly learn this lesson. Their jealous guarding of their power may surface in destructive negativism. Positive reinforcement and the encouragement of initiative are not likely to be part of their mental processes, and just the opposite may be observed. They want "gofers" only.

Manipulators are more subtle. They are as smooth as oil on troubled waters. They may call their people friend, and buddy, and associate. They may name their secretary as their boss, or the one who keeps them up to speed and doing what needs to be done. They also strive to be seen as right, sometimes even when wrong, but especially when neutral or oblivious. They will attempt to evade deserved blame, and try to cover their trail when they know that they have erred. They want scapegoats.

Fussbudgets are meticulous to the point of often losing sight of the stated objective of their efforts. Form is more important than function or results. Good staff work can be discouraged because the fussbudget fails to evaluate its content, being distracted by the format and frills, the missing comma or misspelled word on the production status report. They might fail to gain important information on a timely basis, because their subordinates have learned not to submit anything that is rough, partial, or preliminary.

Commandos are trained to charge, and managers of this ilk will try to be heroes by charging ahead on a project with half of the information needed to make rational intelligent decisions. If they can be the first to succeed at a new system, the first to bring in a new product or machine, they will indeed be heroes, but all too often, they stumble over unknown, unforeseen, or unevaluated hurtles. They are cousins of the tyrant, in that they alone make the decisions. They have an optimism that overshadows intelligence and facts. They want yes-men who will follow them enthusiastically into the unknown.

Loners are likely to be workaholics. They may have competent staff, but they either have to do it all themselves, or they have to verify everything independently on their own. They tend to make all of the decisions, like the tyrant and commando, not because they are selfish or uninformed, but rather because they really don't think anyone is competent to do it for them or to help them. They don't know about, or don't believe, in synergism. They want bodies, but not help.

Obviously, few individuals will exactly match any these stark stereotypes. As in our discussion on management involvement (proactive, reactive, and passive) all shades and combinations exist in the real world. This is also true of the stereotypical fair-haired, intelligent, supercompetent, thinking, planning, doing, professional, likeable, fair, trustworthy, considerate, kind, prioritized, team-oriented, cost-conscious manager. In a less intense embodiment, the dictator may in fact seek and use helpful input and advice, but still make unilateral, or self-enhancing decisions. The fussbudget may be a stickler for final form, but make timely use of current information. Workaholics or loners may not try to do it all by themselves; they may, on a larger scale, try to have their department, as a group, do it all for the whole company. Similarly, no manager, however competent and successful, is always so, and possesses all of those ideal traits we wish they had. Tyrants and commandos are not necessarily failures. But they are not as successful as they could be.

Dealing with Various Philosophies

Safety professionals have three fundamental choices available in dealing with operating managers. First, they can meet them on their own turf. They can attempt to bully the tyrant, or manipulate the manipulator, or overpower the commando. Secondly, they can choose to be subservient and helpless by communicating the problem, and then washing their hands of further responsibility. Third, and obviously correctly, they can relate to the manager in a wholly and continually professional manner.

But what is the professional approach to dealing with a tyrant, a dictator, a bull-of-the-woods, do-it-my-way, mentality. Suppose that the safety professional has just informed a tyrannical operating manager of a hazard or violation, and the manager responds negatively. Our advice is that the safety professional not acquiesce, but not argue or interrupt either. He or she should listen attentively, make notes perhaps, and maybe even paraphrase the manager's statements or ideas, in order to exhibit understanding. Whether the manager's views on safety or a safety problem take the form of a tirade, a threat, derision, simple disagreement, or contempt, the safety professional must truly reach an understanding of what the operating manager thinks, the arguments, valid or otherwise, that an upper level manager might hear if the issue becomes elevated. At the close of such a dispute, the safety professional should thank the manager for his or her thoughts and evaluation, and assure the manager that they will be given full consideration in the final assessment of the matter.

The manager's thoughts and evaluation should then *be* given due consideration. Whatever the manner of presentation, legitimate facts and views must be acknowledged. Similarly, unsupported opinions, emotional likes and dislikes, simple disbelief, or blatant disregard for the problem, cannot be ignored. But neither can they be allowed to be the final word.

The safety professional, at this point, has to accept that his or her professional expertise, knowledge, or assessment is not, by itself, sufficient to penetrate the tyrant/manager's wall of negativism. Each point made by the manager should be taken, and the requirement that applies simply cited. If it is an OSHA regulation, chapter and verse should be cited. If it is an NFPA requirement, likewise. If it is the safety professional's own judgment or observation, the hazard should be described, how it has developed, how it can injure or damage, and how it can be rectified. This is where the plausible economics of injury and corrective action costs can be used very effectively.

The safety professional should present the re-analysis to the offending manager, not as a refutation, but as a factual and supportable documented re-analysis. If the reaction is positive, the case has been won; the hard part is over. However, if the tyrant/manager still is unmoved, professional ethics must prevail, and the matter elevated to higher authority.

Dealing with a manipulator is often easier, because the last thing a manipulator wants is to look bad. He or she will probably respond smoothly and positively to the safety professional's disclosure of a safety concern. The word of warning in this kind of dealing is to follow up to ensure that the manipulator acts, and doesn't just talk.

Dealing with a fussbudget/manager may be a bit distracting. The caution here for the safety professional is to present the problem in well-thought-out, all-bases-covered, detail. He or she should try to anticipate all possible questions, even irrelevant ones, that could arise. The cited source should be handy, and the safety professional ready to describe, or show, the problem to the manager, and be able to recommend a detailed plan for correction.

The commando will probably need restraining, more than convincing. His or her enthusiasm will be welcome, and the safety professional will really have to try to hold the manager down until sure that the manager is planning the right approach.

Dealing with the one-man-show, or with the journeyman, will be similar. Both will probably react in a competent, interested, and concerned manner, and want to alleviate the hazard. If that is so, the safety professional's mission has been accomplished. How the action is taken, and by whom, is not an immediate concern.

In any of these dealings, three basic tenets should be in the forefront of the safety professional's mind. The safety professional should be sure of his or her position concerning the hazard or discrepancy, and its mitigation. There should be no hesitation to elevate the issue to higher authority if the operating manager is balky. Also, the safety professional should always follow up to see that the action really did follow the words.

Interpreting Actions and Decisions

Actions, decisions to act, or failure to act, follow every safety recommendation made. Decisions are made one way or another, by pocket veto, trashing, taking a minimum approach, studying the issue to death, or by careful, economical implementation of the necessary corrective measures. Sometimes, there might be an argument over an interpretation, or whether a certain regulation applies. Some operating managers might emulate Philadelphia lawyers, trying to find a loophole in the regulation, which will excuse them from acting. Most of these responses may be made in the form of a tirade, ridicule, debate, or a rational attempt to avoid an action if possible. Safety auditors are often viewed as bringers of bad news, so they have to be prepared to take the heat. As we have advised before, if the exposure/risk is real, articulated plainly, backed by data or regulation, then the heat will be easily endured. A vague and opinionated description of the hazard will earn justified disdain and inaction from an operating manager who is leaning toward inaction anyway.

One of the early clues for identifying managers with that leaning is when they refer the problem to a committee, or to an ad hoc group to study it, or to their area foreman for evaluation. Another clue is the response that says, in effect, "I'm sure you're right, but I don't know what I can do about it," or "My budget won't stand this," or "Why don't you take this up with the Finance Department."

This kind of evasiveness can be a dead end if safety professionals allow it. However, they can also adroitly elevate the issue with something like, "This problem is in your department, Mr. X, but is also one of Mr. Y's (the manager's boss') departments. I think we should get his views before *we* make a final decision, don't you?" Of course, the same tactic can be used on Y if necessary, and elevate it to Z, the vice president or plant manager. The safety professional needs to be more and more sure and specific each time he or she elevates it, though.

A commando manager might enthusiastically receive a report of a problem and promise to get right on it, and promptly forget it, or worse, take inappropriate measures. A clue to this pitfall is when the operating manager tells the safety person that it will be taken care of, and then (in effect) dismisses the safety person without reaching some agreement on what will be done, and by when. The fussbudget is likely to have a committee study the matter, and make a written report to him or her at a later date. The tyrant might fume a little, but decisive action will occur if he or she is convinced that the problem does indeed require correcting.

Any promissory positive response to the news of a safety problem requires follow-up. The vaguer the promise, the less likely it is that action will actually follow. Safety professionals should try for two commitments from operating managers when informing them of the issue. They should do their best to illustrate the reality of the problem, to convince the manager that action is needed, thus getting a positive commitment to act. Second, they should zero in with the manager on a completion, or abatement, date.

If the operating manager seriously agrees to correct something by a specific date, the odds are he or she will do so, or at least try very hard. If the agreement is vague or grudging, watch out. In any case, follow up should be an integral part of the safety auditor's game plan.

One of the most frustrating aspects of approaching an operating manager about an unsafe condition or work practice is overcoming his or her seemingly genuine disbelief that it is something real and threatening. Selling prevention is tough. The well-worn rebuttal that that accident hasn't happened yet can get all too familiar, and a fairly sure sign that it is going to be a tough convincing job. Historical facts, regulations, and data are the selling tools we keep coming back to, and the cost of accident/cost of correction/probability equation previously described.

Large/Small Contrasts

In keeping with our penchant for oversimplification, we will suppose that there are two kinds of companies, or organizations—big and little. Some will opine that big is bigger than large, and that little is smaller than small, but we'll use these term pairs interchangeably to break the monotony.

Safety in a large company will likely be a structured element of the organization. There will probably be a corporate director of safety, with local safety managers at the several plants. The relationship of the local safety managers to the director may be one of direct subordination, or a dotted line relationship, in which the local manager reports administratively to the plant manager, but receives policy and program direction from the corporate director. In the absence of a corporate director, or equivalent, the plant safety manager will generally answer to the plant manager.

In a big company, the ideal location in the organizational structure for safety auditors is on the corporate staff, where they are free of any pressures from, or vested interest in, the plant or plants that they are auditing. As in any auditing function, distance between the auditor and auditee promotes objectivity, thoroughness, and professionalism. From the corporate staff, probably reporting to the corporate safety director, auditors have access to the boss, they know first-hand the safety policies and posture of the organization, what the standards are, and how they are interpreted. They may even have a hand in policy formulation through their findings and recommendations. When auditing in one of the production departments in one of the plants, representing the corporate office and the safety director promotes a certain aura of respect and credibility for the auditors also. But that is only at the introduction; the auditors obviously have to earn sustained respect and credibility.

We should probably suggest here that it is very difficult for a plant safety manager, or one of the area safety engineers, to conduct a true safety audit. Not impossible, just very difficult. There is the matter of vested interest. Time and time again, we see where safety engineers on the floor become part of the crew. They have bought into whatever exists; they are a party to it. Having nothing to do with the question of professional integrity, self-evaluation and self-criticism are extremely difficult, and virtually impossible to do with the same degree of objectivity that someone else can.

There is also the tired, but true, cliché about familiarity breeding contempt. More correctly, familiarity breeds acceptance, perhaps even obliviousness.

If we put our minds to it, we could probably find a dozen safety problems in our own homes that we as professional safety auditors have become inured to, simply because we see them every day, or perhaps because we see them, but don't *see* them.[1]

For similar reasons, the person who wears the "safety hat" in a little company is probably not the best one to conduct safety audits there. This is not to say that either one should abandon efforts to find and resolve safety problems through inspections and work practice observations, job safety analyses, etc. It simply means that auditors should be from outside the organization that they are auditing.

In a small organization, where probably one person alone wears the official "safety hat," the safety person or the company's management should consider retaining a competent independent consultant to do the audits. The week or month that the consultant would spend evaluating the compliance status and hazard control would be nominal compared to a permanent staff addition, and certainly negligible next to the costs of a catastrophic accident.

Client and Employer Relationships

There are a number of similarities in the relationship between the safety professional and the employer, and his or her relationship with a client. Both have hired the safety professional and will presumably pay for his or her service and expertise. Both can fire the safety professional. Both have some kind of expectation, however

---

[1] *Note:* After writing this section, both authors did a "safety audit" of their respective homes on the same evening. Next morning, Kay came in with seven problems, easily corrected, and Don had nine.

clear, vague, or unrealistic, that somehow their enterprises will be better off for the relationship. Either may be hoping for a panacea; either may be succumbing to political, legal, or economic pressures to improve and change things. Alternatively (fortuitously for the safety professional), either may have a truly enlightened desire to get the expert help to maximize productivity and profitability, and reduce human trauma, by making their operation as safe as it can be.

However, there are also some striking differences in these two kinds of relationships, and in how safety auditors, in particular, should interact in them. In both cases, let's presume that the decision to initiate a safety auditing program is a new one. We have discussed in some depth the safety professional's dealings with his or her own operating management, the kinds of management styles that can be encountered, recognizing, interpreting, reacting and interacting with various kinds of managers. Safety auditors in an employee's role are faced not only with identifying safety concerns and problems, not only with recommending corrective actions, but they then have the task of selling the whole package to the responsible operating management. Ninety-nine percent of what they are selling is prevention, and they are selling it to colleagues in the business. They have a very real and close personal stake in the success of the enterprise, and in their continued association with it. Their livelihood can depend on how they deal with those colleagues, and how upper management views their activities.

As consultants to clients, the personal involvement can be greatly diminished. Of course, safety professionals have an interest in establishing a successful and rewarding (and profitable) relationship, but they do not have the emotional bias that inevitably surfaces in employee/employer relationships. Their contacts in a client company are not colleagues or bosses. They can overlook in-house politics and be totally objective in their statements, without concern that they might be resurrecting old ghosts. They are known outsiders, and are expected to be ignorant of prejudices and traditions. They are in the client's facilities at the latter's overt invitation, and not to be perceived as foisting themselves onto the turf of old friends or colleagues. The freedom is refreshing.

This is not to suggest that a safety auditor employee cannot be just as objective, ethical, and professional as a consultant, nor is it to intimate that a consultant has to be more critical or less empathetic than an employee colleague. Professionalism is a state of mind that adapts to the relationship so as to maintain itself, irrespective of the nature of that relationship.

The professional safety auditor is providing useful, credible and sound information to the operating manager. The content of that information should not be a function of who is being related to. The context of the information *must* be a function of the relationship. The colleague operating manager has probably not asked for the safety auditor's advice. The client has, or he wouldn't be a client. Usefulness, credibility, and technical soundness will probably reduce the difficulty of selling prevention to the client. It is an essential ingredient in selling prevention to the colleague.

All of this seems to be saying that it is easier to audit a client than one's own employer's operations. That is exactly what we are saying. The findings should not vary, but the use of the findings, the way that they are presented, will vary. It is definitely and inherently harder to identify problems to a friend and colleague than to a stranger. Still, objectivity and accuracy must never waiver.

We have developed a self-evaluating checklist for safety professionals and auditors to use to satisfy themselves that they have maintained their objectivity on an audit project. This is given in Table 2.1 Obviously, this checklist will only give valid answers if used objectively.

Table 2.1. Objectivity checklist.

1. Did I describe the situation exactly as observed?
    a. Do my field notes match my memory?
    b. Does my written report match my notes?
2. Did I cite a correct applicable reference (company policy, OSHA, etc.)?
3. Did I inquire enough to fully understand what I observed?
4. Have I treated similar observations in the same way?
5. Have I accorded the appropriate degree of seriousness to the observed condition or practice?
6. Would an impartial but knowledgeable stranger detect any element of rancor, excuse-making, condescension, or accusation in my report?

References

Blake, R. and Mouton, J. 1964. *The Managerial Grid.* Houston, TX:Gulf.
Herzberg, F. 1966. *Work and the Nature of Man.* Cleveland, OH:World.
McGregor, D. 1960. *The Human Side of Enterprise.* New York:McGraw-Hill.

Chapter 3

Successful Auditing

Successful auditing is the whole theme of this book. It would gain no one anything to embark upon a safety auditing program if there were no anticipation of, nor measure of, success. Success of a safety auditing program can only be measured in terms of the change it effects on the overall culture of the operation, and enterprise that it audits.

We will discuss the measures by which operations, its management, and the safety auditor might be attuned to the same goal.

What is operational success? It is the melding of all goals of success into a single philosophy; that safe production is the overall objective — not just timely and cheap production by itself.

This chapter deals with this particularly difficult issue. What are the measures of success, and how does the safety professional influence them?

Operational Success

Success is one of the most elusive concepts, and ambiguous words, in the English language. Defining operational success sometimes defies our ability to articulate. To the CEO, it is probably very heavily weighted by the profit and loss statement. To the office custodian, success might be measured in terms of how much free time is left after he had completed his rounds of the waste baskets and lavatories. To the machinist or equipment operator, it may be simply getting through another paid day on the job, or it may be the satisfaction of having successfully produced a difficult part on the first try.

One description we like, however altruistic it sounds, is that "operational success is the safe, timely, and economical production of quality products." Operating managers who adopt this motto, understand it, and use it as their guiding principle in decision making, will be successful. Of course, it has to be more than a motto. They should never make a decision or take a course of action that doesn't meet all four tests. They should never allow conditions to exist, or things to be done, that will, in the long run, work counter to any one of those measures.

A compromise in quality may very well allow a higher rate of production and probably lower costs. That looks, in the short term, like success. But inevitably, poorer quality will erode product acceptance, reduce demand, result in lower prices and higher costs, and all the other negatives that can be deduced from such a decision.

By the same token, and just as inevitably, an unsafe practice or condition, or the absence of systems and disciplines to identify and correct them, will eventually lead to accidents. It is axiomatic that injuries, property losses, even near-misses, cost money and lost production, divert management attention and activity, and cause some level of human trauma.

So what does operational success look like? How do safety auditors define it so that they can objectively identify where the operation's strengths and shortcomings are? They have to have a standard, a formula. They have to determine (1) whether the operating manager has the formula, and (2) whether it has been made to work. A formula that isn't used will never produce the desired results.

We propose that the formula for operational success is the identification and control of hazards. It's that simple. This is borne out by the idea, an axiom if you will, that the loss of profits, eroding product value, failure to meet commitments of delivery, and losses from accidents are all hazardous to the enterprise. Safety hazards are only one kind of problem that successful operating managers must cope with.

By many standards, managers who get the product shipped on time, have a low scrap or rejection rate, and make a good profit doing so, are the picture of operational success, because they have, in fact, dealt successfully with the economic, quality, and delivery hazards involved. They heed trends and warnings. They make corrective action decisions that will reverse adverse cost or quality trends, or slow-down in productivity—the hazards that they know and see and fend off every day.

One can even extend the analogy to human relations. It is well known that poor relations with peers and subordinates will stifle "hazardous" information flow, and leave the operating manager without all of the weapons needed to combat the productivity hazards that must be contended with.

It is the integration of safety into their management formula for operational success, "the identification and control of hazards," that will enhance their chances of success in safety as well. To illustrate this crucial point, let's pose the question, "what are the benefits of well-trained and proficient workers?" Isn't it clear that well-trained plant operators will produce better parts and products? Shouldn't it follow that if individuals are virtual experts in what they do, they will do it more efficiently, and at a lower unit cost? Would it not be expected that those who know and understand the potential safety hazards of their job will be more likely to do the job safely? We think that the answer to all of these questions is "yes." This is the integration of safety into operational success. The same management action will enhance all four cornerstones of that success.

At the risk of belaboring the point, let's try again with the equipment-machinery, facilities, and tools that those workers use, and which the operating management provides. Is it reasonable to conclude that efficient, well-maintained equipment will operate more consistently (quality control), reliably (schedule and cost control), and failure-free (safety control)? We think so. A comprehensive preventive maintenance and upgrading program again benefits all of the measurable elements of operational success.

Managers faced with a shortage of raw materials will not be successful for very long if they sit back and say, "Gosh, I hope Purchasing finds some more somewhere." Or when his scrap starts to climb, he thinks "I hope they turn that around pretty soon." Successful operating managers have in place, the tools, systems, and sequences to identify rising problems, meet them head on, and solve them. They shouldn't, therefore, be told of a rising number of injuries and say, "Gee, let's hope people work safer." They won't get much response if they put up posters saying "USE SAFE MACHINES," or "USE EFFICIENT TOOLS" when safe and efficient machines and tools haven't also been provided. Neither will it be effective to exhort people to "WORK SAFELY" if they haven't provided the training, procedures, tools, and equipment to do so. (We are not anti-poster or anti-slogan. We simply do not accept them as any kind of a cure-all for management problems, or as a substitute for management involvement.)

Integrated safety, cost, quality, and schedule management *is* the essence of operational success. Why not then manage them all in one Integrated Management Control System. Too often, we see a Cost Reporting and Control System, a Master Scheduling System, a Quality Monitoring System, and, when we're lucky, a Safety Program. Is there any real reason why these fragments of the whole, often administered by different organizations, should be separate? We will discuss this topic in greater detail later.

The Successful Manager

For operating managers taking advantage of the material here, we have tried in the preceding section to describe what success looks like. For safety professionals who are interfacing with operating managers, we have tried to channel their thinking beyond the visible and material things that they can see — to take a picture of, and into, the realm of objectives and ideals. Safety auditors have many opportunities to observe not only what is, but also infer *why* it is, what things are important, and those that are not. In addition, they may be able to, and should try, to steer managers into a proactive role in safety, to show them how they might actually begin to control their own destiny, regarding accidents in their departments.

If successful managers are dedicated to the safe, timely, and economical production of quality products, what are the clues to that dedication? How can the observer, the safety auditor, identify that *total* dedication? In Chapter 1, we discussed the interrelationships and distinctions of responsibility, authority and accountability, and also proactive, reactive, and passive management. We also stressed the importance of sizing up managers to ascertain which prototypes they most resembled, and how to deal with them. We presented a list of things that we felt safety-motivated managers would do to ensure the safety of their operations (Table 1.1). These activities are observable.

However, successful operating managers often have an almost

tangible aura that bespeaks their success. Some of their observable characteristics are likely to be: confidence, knowledge, energy, enthusiasm. They will exhibit confidence in their own and their subordinates' capabilities. They understand the intricacies of their operations, including the hazards involved and how to cope with them. They are people builders, facilitators of the individual's growth, excellence, and success.

They can be seen on the floor, discussing issues, asking questions, learning and teaching, offering advice. They probably take notes on things to follow up on, including and especially the things that might interfere with, slow down, or otherwise disrupt safe production. They rarely blame others or circumstances. They take responsibility and hold themselves and their staffs accountable. They are proactive. When a solution is larger than they can handle, whether for technical, budgetary, space, facilities, or any other reason, they plan their strategy, prepare their case, including the payoff, and charge off to enlist whatever aid they need.

They not only demonstrate, but believe that their job is to facilitate safe, timely, and economical production of quality products. Could passive tyrants be expected to look like the model we have described? Could reactive manipulators? How about proactive fussbudgets? Passive tyrant managers will throw their tantrums from their offices, blaming any and all in their way, including, no doubt, the safety auditor. Reactive manipulators may respond to a hazard identification by covering it up. Proactive fussbudgets run the risk of using all of their energy to swat the gnats, oblivious to the alligators.

Operating managers that aim for excellence and success, as distinguished from just getting a job done, will work hard to maximize their chances of achieving and maintaining success. If they are so motivated, safety will be a part and parcel of their concerns and goals, and their way of managing.

The Successful Safety Auditor

We have talked about successful operations and their successful managers. What can we offer about successful safety auditors? Clearly, they have to relate to both of the above, whether successful or not.

Successful operations are those that are producing at or above expectations, and are in sync with top management's wishes for those operations. Successful managers are those who have made that to happen, or who have inherited a situation that fosters that kind of success-orientation.

Safety auditors, whether a part of, or a more distant reflection of, the operational scene, are obliged to "tell it like it is." They should have no allegiance to anyone, other than their consciences, as to what and whether they report safety deficiencies. Their entire focus and purpose should be concentrated on what the existent discrepancies are, and what can be done to correct them.

In doing so, they must remember that their purpose in life is to maximize the success of the operations that are under their scrutiny. To that end, successful safety auditors do not attempt to tear down the integrity or dedication of the operating management, but rather build up the nascent commitment of that operating management to operational excellence. The better they can portray the management, the more likely they are to fulfill those expectations.

The credentials for successful safety auditors are very nebulous. They may come from the ranks of those who have many years of experience in the technology of the operations under audit. They may be people who are extremely well versed in the statutory requirements of the particular industrial operation in question. They may, however, also be people who are novices to that particular technology, but are very well indoctrinated in the existent rules and regulations that govern such operations.

Whatever, their background and "credentials," it is important that safety auditors not be perceived as being wholly ignorant of the technological principles involved in a given operational enterprise. Credibility, as we have alluded to before, is one of the safety auditors' most precious allies. Successful safety auditors are those who have studied the operation to the point of total understanding, and interpreted such in terms of its relevance to the every-day conduct of that operation.

Safety Auditing — A Management Training Tool

We touched on the concept of safety professionals as potential trainers and educators. They do, in fact, have the opportunity to instill safety consciousness and proactivity in the operating people and management that they deal with. Often, managers have come up from engineering or business backgrounds, or have come up through the ranks because they are knowledgeable and hard-working individuals who stand out in the department. They are intimately familiar with operations and all of the nuances and subtleties that can aberrate them. However, they probably have not been trained in safety regulations, hazards identification, or how to mitigate them.

Safety auditors can strike a responsive chord with receptive managers if they use their findings to instruct, rather than to accuse or denounce. The latter will usually arouse antagonism, and make the process of auditing more contentious. A neutral, factual recitation of safety problems will likely result in corrective actions that address the symptoms—the individual (perhaps relatively innocuous) findings, and go no further. However, an instructive, constructive illumination of the deficiency (vis-à-vis the requirement), of the hazard, the risks, and the way to control them, can stimulate an operating manager's own inventiveness and intelligence in devising effective and broader corrective measures.

This approach will, of course, be most fruitful with managers who approximate the description that we gave in the foregoing section. It will also be effective with managers who have not yet integrated safety into their management style, but are proactive problem solvers at heart. Managers who rile at criticism may also rile at having a deficiency identified, however constructively done; but he is less likely to. There are also those who will react negatively, no matter how softly the problem is disclosed.

To illustrate, suppose a safety auditor determines that the emergency egress from a work station is inadequate for the hazard and the occupancy. The problem can be identified thusly: "Hey, Bob! Why in the heck did you put those cabinets in the aisle? Don't you know that you have to have a 44-inch-wide escape route? What are you trying to do, trap those guys up there?" How kindly do you suppose Bob will react to this "finding"?

A neutral report will say "Life Safety Code (NFPA 101) requires 44-inch minimum exit access; minimum observed width is 31.7 inches." Bob will probably move the cabinets somewhere else and forget it.

An educational approach might go something like this: "Bob, do you understand that this work station is a hazardous location because of the flammable solvents used there? The life Safety Code is based on observed egress rates and provides guidelines for escape route parameters that will maximize the chances of all personnel safely evacuating in an emergency. If we can open up those aisles to 44 inches, you will be in compliance, and your crew will be able to evacuate quickly if a fire, or any other emergency, should occur."

These starkly contrasting tones are of course contrived to make a point. The first monologue is accusative and will be seen as such. The last one may sound like Pollyanna, but at least it identifies the hazard and the risk of injury, and suggests a way to correct it. The manager can understand (not necessarily agree with) the safety auditor's point, and recognize that noncompliance is something to be avoided. If there is a leaning toward proaction, rather than reaction, the manager may even take this new knowledge, and find out whether there are other restricted egress routes.

A successful safety audit is one in which problems are understood and communicated, and in which both auditors and auditees have learned something.

Successful Communication With Management

We suggest that there are three levels of interest that operating management might have in safety for its own sake. These are economic, compliance, and humanitarian. The bottom line is a common denominator for successful managers, and its enhancement is a goal even for the less successful. Black and white, plausible data showing the economic payout of safety efforts are an easy sell; communication of the data will sell the project.

There are several ways in which safety professionals accumulate and present such data, and they have their places. A simple approach is the totalling up of the direct costs of accidents in terms of dispensary and hospital costs, doctor's and other professional's fees, medical supplies, replacement or book costs of equipment or product damage, wages for compensatory time off, and so forth. The manager sees this data month-in, month-out, year-in, year-out, and it may begin to seem like the property taxes or the fuel bill. It is there, it costs, it varies somewhat around some midpoint, but it doesn't seem to the manager that there is much that can be done about it.

Safety professionals may even go a step farther in trying to stimulate managers. They may estimate the lost production value due to downtime when there is property loss or damage. They may toss in the overtime, retraining, and other personnel costs to fill the slot vacated by an injured employee. They may even put a dollar value on the supervisory and engineering time expended to correct an accident cause, and the material or equipment outlays that may entail. They now have an even larger number for these costs, and may (probably will) get more management attention focused on the safety problems.

Another ploy that is sometimes effective in communicating successfully with management, in order to stimulate proactive safety, is to relate the costs of accidents and injuries to the added volume of production and sales it took to recoup those losses against profits. It is fairly simple arithmetic, if you will bear with us. For example:

Sales	$1000 (add zeros to suit)
Net Profit	100
Accident Costs	50

The profit without having had any accidents would have been $150. To make $150 *under current conditions*, sales will have to blossom up to $1500. Which is easier for management to tackle, increasing sales by half, or eliminating the accidents? Aggressive, proactive managers may try both, and why not?

Once economically driven managers see some options, and see data that they can work with, they will be more receptive to the proactive safety role we are trying to promote for them. The same imaginative talents that they bring to energy conservation, purchasing, and scrap control can also be channeled into loss and accident reduction too. We have successfully communicated the challenge.

Managers who fear noncompliance are ripe for educating also. They may have already endured some level of pain for noncompliance, or they may have read very unpleasant news stories about fines and even imprisonments. To operating managers who are not yet aware of the dangers of noncompliance, safety professionals can tactfully point out in a non-threatening way, that this is today's world, and that they (the managers) may have exposures worth correcting. Safety professionals should quickly add that they will gladly help, emphasizing "help," identify those exposures and the necessary corrections.

Whether driven by fear, a conviction that compliance is just good business sense, or the belief that that which is in compliance is *SAFE*, operating managers who strive for compliance can likely be educated as to what compliance looks like. They are ready for successful communication. They can be taught to develop self-generated compliance. Safety professionals can be the yardsticks and helpers in continuing to communicate the hows and whens and whys of compliance.

Successfully communicating with caring managers, the humanitarians, is of course a relatively easy and rewarding thing. If, in addition, they are bottom-line oriented, they will disdain frivolous niceties, as should loyal safety professionals. Neither will be around very long if profitability is neglected or squandered. Caring managers will not only strive for profitability and compliance, but will do so in a manner that communicates a sincere desire to make their operations pleasant, safe, and interesting places to work. Safety professionals or auditors probably need only to identify ways to avoid accidents and noncompliant situations, to stimulate action in them. Correcting errors, providing equipment improvements, additional lighting, insisting on purging deteriorated tools (or purging such tools themselves), are things that proactive managers will do when they see the need. In addition, the positive impact on their staffs will be quite remarkable.

Regardless of what the operating managers' motivations are, safety auditors must always remember that their job is to help the managers, and their operations to be the most successful that they can be. The operations' success is the safety auditors' success. How they promote that success will be dependent on the kind of people that they deal with, but the focus of their efforts remain the same.

Chapter 4

Involving Line Management

Getting line management involved (getting them to buy into the responsibility for, and solution to, safety problems) is, without a doubt, the largest problem that safety auditors face. In the conventional industrial world, the identification of safety issues carries with it job description requirements that safety professionals "correct" them. Usually, no one can answer the attendant question of how they (the safety professionals) are supposed to do this.

They normally have no authority or budget to make the right things happen. Their dictums usually consist of "ensuring proper compliance with applicable federal, state, and local laws and regulations," and "correcting safety discrepancies where found." Obviously implied, and (lest we beat it into the ground) without merit, is the proposition that safety professionals can effect the overtly stated, albeit shallow, objectives elucidated by management in such decrees.

Unfortunately, this is the normal state of affairs for safety auditors. They might very well be expected to "identify and correct" all safety discrepancies discovered. That this is an impossible "no-win" situation has been, and will be, driven home repeatedly in this book. It is *only* through the involvement of line management, and their acceptance of responsibility for "what is," that progress toward "what should be" will be made.

Educating Management

We have dwelt at some length on the types and styles of managers, the prototypical, stereotypical kinds of people that a safety auditor may encounter, and how to recognize and deal with them. Regardless of the types of individual that the safety auditor interfaces with, (the auditor) must never lose sight of his or her own mission, which is to maximize the success of those managers and their operations. There is no room in the professional protocol for antagonism, retribution, accusation, or exaggeration. The safety auditor must educate operating management through the audit findings, and education implies convincing, which virtually requires cordiality.

Someone, at sometime, decided that safety auditing was right for his or her operation(s). It was possibly a plant manager, a corporate CEO, a corporate safety director—perhaps even the safety professional. If it was the latter, he or she undoubtedly sold the idea to someone in the upper echelons, and got the go-ahead to audit. In our case, it was a plant general manager, who wanted a staff dedicated to the auditing task, and free of distractions that would dilute the effort.

Wherever that decision/approval originated, or will originate, the safety auditor should begin by educating that person. The auditor should first assimilate the assignment, as developed in Part II of this book, and reach his or her own conclusions as to how to go about the auditing. A lot will depend on the nature of the assignment, of course, whether it's a client or employer, whether it is temporary or "permanent."

The auditor should outline the approach, protocols, and schedule to be used, and what kinds of findings he or she can expect to report. It is important that the assigner understand what the assignee is going to do, how it is going to be done, when, and how long it will take. It is also important that the assigner understand the yardsticks, the standards against which the audit will be conducted, and that some are mandatory, while others are advisory and still others may be a matter of professional judgment. The assigner, client, or employer/approver has to buy into the audit plan, approach, and protocols before the safety auditor can launch the program. Nothing is more demoralizing and credibility-shattering than to have the instigator of an audit undercut the auditor and the findings with, "Why did you do it that way?" or, "This isn't what I had in mind at all."

An important feature of this educational interchange is to assure and convince the assigner that the audits will be impartial, and that the auditor will report areas of excellence along with deficiencies and problems. In this introductory briefing (which may have more than one iteration), questions, comments, and possibly direction are likely, and the plan should be modified to accommodate the wishes of the assigner. On the other side of that coin, the auditor should not allow the modifications to encroach on professionalism from either direction. Neither watered down problem identification or excuses, nor punitive reporting of deficiencies, should be accepted as the way of conducting safety audits. The safety auditors must convince his or her superior(s) that facts versus requirements is the whole name of the game.

Once an agreement is reached, the auditing program has become the assigner's, not the assignee/auditor's. This is important in the next step of the management education process. This step is the introduction of the auditing program to the operating manager(s). We recommend that before this is attempted, that novice safety auditor create a document outlining the audit program, the purposes, the scopes and protocols, and the handling of findings. It should be in a recognizable format within the management document system of the establishment, and may be variously known generically as a Management Policy, Plant Directive, Executive Order, and so forth. The points of the agreement reached with the assigner should be couched in the normal jargon of such management directives, then approved and signed by the executive who normally does so. In the case of a client/consultant agreement, the contract should reiterate much the same points, and will, of course, be signed by both parties.

Now the safety auditor has the management backing needed to enter the operating areas and conduct safety audits.

It is also our recommendation that an auditor allow sufficient time to elapse so that he or she is reasonably sure that the operating manager is aware of the audit program through dissemination of the management directive that established it. It is far less intimidating to the manager when the auditor calls and asks whether he or she has seen Management Policy No. XXXX, than if an auditor flashes a copy with the signature still wet. Since it is in human nature to at least be dubious of, if not outright resist and resent the new and different, giving the manager time to assimilate the fact of the audit program will soften the impact when the auditor makes the initial contact.

When that initial contact is made, the safety auditor is faced with a second educating task. It should be made clear that his or her efforts are aimed toward improvement and increased success, that he or she is fulfilling the direction of Management Policy No. XXXX, and that the modus operandi will be a comparison of what is with what should be. He or she should solicit indoctrination and education from the operating manager. Inquiries about the nature of the operation, what the departmental rules are on personal protective equipment, union participation (if appropriate), working hours, points of contact, etc. are all helpful. The scope and depth should be outlined, as well as how the findings will be reported. He or she should, at the same time, offer informal briefings at the operating manager's convenience and request.

All of this is not intended to be solicitousness for its own sake, nor to convey an approach of subservience, but merely to show that, while on the operating manager's turf, the safety auditor is a visitor with a purpose and a plan. The auditor should invite accompaniment by the manager or one of his staff, but not demand it.

Finally, having done his or her homework, the safety auditor should enumerate the standards (what should be) that will be audited to, and make sure that the manager understands and accepts the applicability of each and every one of those standards. Any questions or disagreements on this point should be thrashed out right then. The auditor should take the time to explain why the standard, regulation, company directive, etc. applies, listen to questions or arguments as to why it does not, and finally agree or arbitrate, as necessary. It is worth the time up front to educate the operating manager as to what the laws, codes, and management have decreed. The manager will be much more receptive to a factual reporting of findings, and to initiating corrective action, when he or she is so educated. Once again, the auditor should tell the manager, and then deliver on the promise, that he or she will report compliance with those standards, as well as noncompliance.

Interviewing the Operating Manager

As we noted in the preceding section, the opening interview, and any followups or continuations, are likely to be an educational process, and can and should work both ways. The safety auditor will want to learn about the management systems and operations, and

50 / I MANAGEMENT EXPECTATIONS

the operating manager will be interested in what he or she is in for, with regard to the safety audit that has been thrust upon him. We have developed an operating manager's interview checklist, shown in Table 4.1. There are twenty-three items that the safety auditor should touch on, when applicable. The checklist is developed with the presumption that the interview is the first meeting between the auditor and the operating manager, and that this is the first audit of the operation. Obviously, if the individuals are old friends or acquaintances, or if the audit is a followup or repeat, not all of the items are appropriate. A supplement for followups is given in Table 4.2.

The Interview Checklist

We would like to offer a brief comment on each of the items in the checklist. We strongly suggest that the auditor make notes or tape the interview, if the manager doesn't object.

Table 4.1. Interview Checklist for Operating Managers.

1. Introduction and pleasantries.
2. Cite management directive(s).
3. State your commitment to objectivity, to improve success.
4. Emphasize "what is" versus "what should be".
5. Cite your background and credentials.
6. Establish the manager's work history with company/operation.
7. Ask—What can you tell me about the technology of the operation?
8. Ask—What can you tell me about the hazards of the operation?
9. Ask—What are the challenges to the success of the operation?
10. Ask—What are the PPE requirements for visitors?
11. Ask—What are the escort requirements for visitors?
12. Ask—What are the normal working hours?
13. Invite participation.
14. Ask—Who should be a point of contact?
15. Describe the scope of the audit, and the approximate duration.
16. Enumerate who will be receiving the report.
17. Offer informal and verbal briefings daily or as desired.
18. Enumerate reference standards (what should be).
19. Explain why they apply.
20. Reiterate the intent to report compliance and excellence also.
21. Ask—What is the manager's role in safety, if any?
22. Ask—Does the manager have a broader role in loss control in general?
23. State when you will begin—allow time for assimilation.

Table 4.2. Interview Checklist for Operating Managers — Supplement for Followup or Repeat Audits.

1. Ask — Are there any new processes, procedures, or equipment?
2. Ask — Have they worked out well?
3. Ask — What is the nature of the changes since the last audit?
4. Ask — Have they created any new safety problems?
5. Ask — What progress has been made in correcting problems from the last audit?
6. Ask — Is the manager satisfied with that progress?
7. Ask — What has impeded prompt correction?
8. Ask — Has the manager's perception of his or her safety role changed any?

Introduction. It is just as important for the safety auditor to establish cordiality and cooperation, as for a salesman trying to land a new account. The introductions, the cup of coffee (if offered), the discussion of weather, business, golf, the family picture on the manager's desk are not inappropriate, if the climate seems acceptable for such pleasantries.

Authority. At some point fairly early on, the auditor or the manager will indicate that it is time to get down to the purpose of the appointment, and the auditor should provide a copy of the management directive under which the safety audit is to be conducted. It should be a rather prosaic statement that the auditor is planning to audit this operation as provided, or directed, in Management Policy No. XXXX.

Commitment. As a part of the introductory statement, the auditor should state, as convincingly as possible, his or her commitment to objectivity, that he or she will not be "out to get" anyone, or to promote pet projects. The auditor should state that operational success is the auditor's goal, just as it must be the goal of the operating manager.

What Is Versus What Should Be. This (to some, trite) summary is a phrase we believe should be repeated often — before, during, and after an audit. It is most especially important in an opening interview.

The Auditor's Credentials. This is the auditor's opportunity to establish an initial credibility. He or she should recite, without embellishment or self-aggrandizement, his or her educational and professional credentials. A business card can be offered. Any professional certifications (i.e., CSP) should be stated. The auditor should briefly describe his or her work history, and (in slightly more detail) his or her record with the company.

The Manager's Background. Inquire about the operating manager's tenure and experience with the company. Do not respond judgmentally as this background is recounted. Merely absorb it.

Technology. If the safety auditor has a technical background, this can be very instructive. If he or she does not, the manager should provide some terminology that will be useful. This question can become an interchange, and the auditor should ask questions, either technical, or explanatory, or both.

Hazards. The discussion on technology can and should lead directly into an interchange on the hazards of the operation. This can be quite informative to the auditor, in determining how well, if at all, the manager has integrated safety into his or her management activities.

Challenges. This is a good summarizing question to recap the technical and safety/hazards discussions.

PPE Requirements. The discussions above may have touched on personal protective equipment (PPE) requirements. The auditor then establishes precisely what PPE will be needed as he or she is on the floor auditing. In some instances, a physical examination or some specialized training may even be needed before the auditor is allowed into an operating area.

Escorts. In some operating areas, visitors may be required to be escorted by an employee or supervisor from that area. The auditor should be very willing to comply with this and PPE requirements and, in fact, should be leery of unsolicited waivers.

Hours. Shift changes, break and lunch periods, and other routines should be determined. The auditor will want to observe other shifts in multishift operations, and the shift turnover disciplines.

Participation. This is a routine invitation to the operating manager to accompany the auditor at any time, or all of the time. It will be rare indeed that the invitation is accepted in part or in whole, but we recommend that it be made every time. When it is accepted, at least in part, the auditor should express his or her appreciation of the manager's interest.

A Point of Contact. The auditor should establish a primary point of contact, the individual from whom data, paperwork, records, drawings, etc. can be requested, and to whom daily or periodic informal findings can be reported. This may, in fact, be the manager, but if not, it should be someone of the manager's choice.

Scope and Duration. If the scope and duration have been previously discussed, it is still appropriate to restate them. If the operating manager has several operations or areas of responsibility, the auditor should point out which one he or she will be concentrating on. If a total, comprehensive audit is planned, the topics that the audit will cover should be reviewed briefly. Also, a target time limit for concluding both the field work and the report should be set.

Report Distribution. The manager should understand up front who will receive copies of the audit report. This will include the manager, of course, and the safety auditor's management, and, depending on the protocols established in the management directive under which he or she is working, various upper levels of management, probably including the manager's own boss.

Briefings. Once again, offering to brief the manager regularly is a very positive strategy. The auditor should try to insist on some periodic review of findings. It should be pointed out that this will allow the operating manager to take some corrective actions before the audit is concluded, and that such actions will be acknowledged in the report.

Standards. Nothing deserves more time then identifying the standards against which the operation will be audited. This will include company and government regulatory standards, and contractual standards if applicable.

Explanation of Standards. This may or may not be necessary, depending on how the manager responds when the standards are itemized. But the auditor should be prepared to explain in a nonemotional way why and how they apply, and if necessary, what the standards call for in the operation.

Compliance and Excellence. The auditor should state again how he or she will note and report compliance and excellence — the extra mile above and beyond mere compliance.

The Manager's Role. Asking a manager what his or her role is in safety matters may catch the manager off-guard. If the manager has never contemplated the question before, we suggest dropping the matter temporarily and pursuing it later in the briefings. But a stated answer to the question will be enlightening to the safety auditor, in understanding how proactive and involved the manager is.

Loss Control Role. This is a corollary to the last item, a potential followup, if it seems appropriate. It can help a manager articulate a perceived role, even if it has never been defined in terms of safety per se.

Assimilation. We recommend that the auditor not proceed directly from this interview to the operating floor to start auditing. Allowing the manager to reflect a bit on the interchange will probably make the next meeting go smoother. It will allow both the manager and the auditor to digest what was learned, and to allow new questions or information to be formulated and presented. Starting one or a few days later would not be wasteful.

As can be seen, a thorough opening interview can be extremely informative, and crucial in setting the stage for a productive and cooperative safety audit. More important, however, than the facts and figures that the auditor obtains from the interview, is the relationship that is launched and molded. If it concludes in an atmosphere of mutual respect, trust, and a spirit of cooperation and constructive intent, then the findings from the audit will be useful and beneficial. If the interview becomes argumentative, offensive, or defensive, or if one or the other party tends toward threat, abusiveness, or accusation, then little beneficial result can be expected. It is incumbent on the safety auditor to do everything he or she possibly can to conduct the interview, and complete the applicable portions of the checklist, in a low-key, educational, fact-finding, and nonthreatening manner.

Tailoring the Interview

In Chapter 2, we stereotyped the management styles that a safety professional may encounter in dealings with operating management. In our admittedly earthy view, we talked about the tyrant, the manipulator, the fussbudget, the commando, the loner, and finally, bless him, the proactive competent manager—the one who truly wants to run a successful operation. We thought that now it would be useful to discuss tailoring the opening interview to these various kinds of individuals.

The introduction will be an early, if not the first, clue that the safety auditor will have as to what kind of manager he or she is dealing with. A tyrant will try to dominate the conversation, perhaps telling the auditor what he or she may and may not do, outlining "rules of the game," perhaps belligerently asking what the auditor thinks he or she can tell the manager about his or her operation. A manipulator will probably try to explain away deficiencies already known to exist. A fussbudget will get all wrapped up in form and format, worrying about each move that the auditor may make, trying to orchestrate the whole project in advance. A commando will likely be ebullient and ready to tell the auditor all that he or she knows and all about what is being done, how the operation is run, and then hope perhaps that the auditor will go somewhere else. The loner may be withdrawn, volunteering little or nothing. The competent proactive manager will be interested in the auditor as a person, as a professional with a job to do, and as an asset toward improving the success of his or her operation.

The safety auditor may have to remind the overaggressive manager that there is limited time and some important information that each must assimilate and understand. To the manager who doesn't want to quit ranting or rambling, a firm interjection such as, "Do you know why I am here?" may be in order. It will lead directly to citation of the obviously needed management directive that the safety auditor must have.

We acknowledge that firmness is sometimes difficult to muster by a safety professional interfacing with a potentially antagonistic manager. The key step is this situation is to avoid returning antagonism with counterantagonism. A more dogged demeanor, returning to the purpose of the interview, will ultimately be more productive and preserve the auditor's professionalism.

The apologies and excuses of the manipulator should be politely heard, then ignored. The facts of the audit findings will speak for themselves; there is no need for the auditor to rely on these introductory attempts at whitewashing the situation. An overly zealous defense of the manager's integrity should be a signal to the auditor that there are probably problems lying unresolved on the floor. Again, the auditor is not fact-finding concerning the state of safety affairs at this point. He or she is trying to conduct a two-way educational interview.

Appealing to a fussbudget with questions concerning the challenges, technical bases, and significant hazards of the operation may serve to get him or her off of the minutiae that this type of manager will try to retreat to. The fussbudget may, on the other hand, dwell at length and in great detail on the PPE and escort requirements, working hours, and protocols. Patience is advised in this interchange.

The commando may need a bit of restraint to get through the conversational details. This manager may be impatient in the portions of the interview that are devoted to policies, credentials, standards, and protocols, and may be quite effusive about the challenges and describing his or her own perception of the manager's role in safety and loss control.

It may take a much shorter time to interview the introverted loner, since the auditor may expect virtually monosyllabic answers to questions, and little or no response or comment about statements.

The safety auditor who encounters a dedicated, proactive manager who is success-oriented will have little trouble or diversion in conducting the interview in a mutually educational manner. The auditor can expect questions, and can anticipate honest answers. He or she should reciprocate.

The bottom line to tailoring this opening interview is not in tailoring the questions and conversation, nor the information sought or given. The tailoring is in being prepared to listen to those thoughts and responses that the manager wishes to discuss at length, and to ask probing follow-up questions, or to repeat statements for emphasis, that the manager feels like dismissing or avoiding. While employing patience when the manager is longwinded, and acting interested even when the auditor is learning more than he or she ever wanted to, a rapport will be built, a sense of relationship, that inflexibility would never permit. But the flexibility stops short of compromising obtaining the facts needed, and does not include failing to establish the legitimacy of the audit and what will be done with the findings.

Some do's and don'ts, depending on the opening introductions and exchange of pleasantries, and what the auditor concludes from them, are offered in Table 4.3.

Table 4.3 Guidelines for Dealing with Different Management Personalities.

For the Tyrant, do:	Do not:
1. Be patient when he explains that he or she is the boss.	1. Argue about the wisdom or effectiveness of safety audits.
2. Try to stress, tactfully, authorizing policies and your commitment to *the manager's* success.	2. Question (at this point) work rules or visitor requirements.
3. Emphasize your objectivity vis-à-vis the standards	3. Respond defensively to a challenge to the applicability of standards.
4. Offer real-time briefings.	
For the Manipulator, do:	Do not:
1. Make mental note of any hasty waivers of rules on your behalf.	1. Accept such waivers.
2. Note deficiencies offered and stated reasons.	2. Be diverted from completing the checklist by the manager's deploring of circumstances.
3. Emphasize your intent to root our factual support for those reasons, to help him or her overcome them.	3. Challenge the reasons given for deficiencies' existence.

(continued)

Table 4.3 Guidelines for Dealing with Different Management Personalities. *(Continued)*

For the Fussbudget, do:	Do not:
1. Be patient in reviewing form, format, schedule, briefings, report distribution, and other minutiae in detail.	1. Allow minutiae to divert you from completing the interview checklist.
2. Follow-up, probing questions about details of the technology, hazards, or challenges.	2. Show disdain if he or she is not as knowledgeable technically as you expected.

For the Commando, do:	Do not:
1. Be firm about completing the interview checklist.	1. Permit him or her to take over the audit.
2. Accept the managers's participation, if the offer is accepted.	2. Accept, at face value, all of the rosy pictures that may be painted.
3. Make notes of all technical details offered.	3. Dispute those rosy pictures.

For the Loner, do:	Do not:
1. Probe for definitive answers, using followup questions as needed.	1. Allow reclusiveness to stifle your completion of the interview checklist.
2. Ask whether he or she has any questions after you have recited your management policy, standards, report distribution, etc.	2. Abandon a point until you both acknowledge an understanding of it.

For the Proactive Manager, do:	Do not:
1. Express appreciation and admiration for his or her attitude.	1. Allow the manager's and your confidence to tempt you to short-cut the interview.
2. State that, as operations confirm it, his or her proactivity will be reported.	

This section has been a thumbnail course in productive human relations, and many others are far more expert in that field than we. Our attempt here is merely to help the safety auditor anticipate and be prepared to stick to the mission he or she has in conducting this opening interview with the operating manager, who hasn't really asked for the blessings of a safety audit.

Coaching and Improving

Involving operating management throughout the course of an audit is an important protocol for the safety auditor. It will ultimately be a measure of the success of the audit and auditing program. It must be clear, however, that how the audit findings are presented, and when, is even more important, because it will influence the manager/auditor relationship, regardless of the degree to which the manager becomes involved in the audit itself. The safety auditor will have defined the authority under which he or she functions, the standards from which, and to which the audit will be conducted, the methods and protocols, and should have reached an understanding with the manager as to both of their roles in the initial interview. As findings are identified and recorded, or as questions arise for which there are no obvious or immediate answers, it is incumbent on the safety auditor to communicate them to the manager as quickly as possible. Alternatively, if so agreed earlier, the auditor should inform the manager's designated point of contact.

If management policies or directives are clear enough, there should be no question that the operating management has the responsibility (and hopefully, the authority and accountability) to evaluate and correct safety problems so communicated. Where that ideal has not yet been approximated, the safety auditor will indeed encounter his or her own problems of communication. For the manager who does not accept, or believe, or understand his or her role, and who questions his or her own ability or authority to effect change, the reaction to findings is probably going to be defensiveness, excuses, and maybe anger. As we have previously discussed, the safety auditor must never return such treatment in kind. Also, the auditor must never shoulder the operating manager's responsibility for correction. The auditor's job should be safety auditing, and reporting to operating and upper management on the findings. The best that can be done is to report those findings in real time, and in a manner that will minimize the frustrations that the less dynamic or effective managers might experience. The manager deserves real-time information so that (1) he or she is not surprised when the audit report is issued, and (2) he or she can at least have the opportunity to plan, and perhaps even accomplish, corrective measures during the remainder of the audit. The manager should know that such efforts will be acknowledged in that report.

It is vitally important that the safety auditor not coach or correct line workers. The auditor should feel free to ask questions about what they are doing; to make inquiries about whether they know and understand the hazards involved; how they do what they do; what they watch for; when; what tools or equipment are necessary; etc. We feel that it is very important that communication be open between the line operators and the safety auditor for purposes of educating the latter, but not for instructing the former. In some operations, the manager, or his or her designee, will demand to accompany the safety auditor during these interchanges, and management's right to do so must be both acknowledged and appreciated. But even if not required, the auditor must not coach, criticize, or otherwise comment on the propriety or correctness of what is observed and learned, except to representatives of management.

Alone or escorted, the safety auditor should not modify, in any way, how he or she interfaces with the line operators, because in neither situation is the auditor "at home." To all in that operating department, and especially to the line operators, the safety auditor is an outsider, possibly someone to be suspicious of. The auditor will have to establish, both in words and actions, that he or she has, and presumes, no authority over the operation, and directs nothing. Line management must direct improvements and changes, and make them stick, and the auditor's only mission is to show and convince them where those improvements and changes are needed.

So coaching and informing is to be used as feedback to operating management alone. The tenor of the feedback is just as important as the specific information itself. We recommend that several basic thoughts be kept in mind in these daily, or periodic debriefings:

1. Always try to present a mix of positive and negative findings (assuming that there are some negative ones).
2. Always cite the source of the requirement or standard (what should be), against which the finding contrasts (what is).
3. Explain "what should be" until it is clearly understood by the manager.
4. Do not hand the manager a problem for which there is no solution. Decisions of this magnitude have to be made at higher levels.

5. Always offer a solution, a corrective action, or a recommendation.
6. Make it clear that your recommendation is probably not the only viable one. But if it is, show why it is.
7. Where options exist, focus on the requirement, and appeal to the manager's expertise to come up with the optimum corrective action.
8. Give the manager time for evaluation and decision making.
9. Make appointments for these debriefings or at least try to catch the manager at slack times.

The safety auditor's "product" is findings — information for management on conditions, practices and procedures, compliance, and management systems that have an influence on the safety of the operation, and that need change, correction, or improvement. Like any other potentially distasteful, but needed product, how they are packaged and presented has an immeasurable effect on how they are accepted and used by the management, whom the auditor is truly trying to help.

PART II

Planning and Preparation

At the beginning of Part 1, on Management Expectations, we stated that the single thing that management *wants* from safety auditing is help, not trouble, and that what they might *expect* is not controllable, at least initially, by safety auditors. Further, what they *get* is totally the responsibility of, controllable by, the auditors. The roughest start to operating manager/safety auditor relationships can become smooth and cooperative ones if auditors maintain their professionalism and produce meaningful results in a nonthreatening way.

A crucial element of an audit, and one that will build an auditor's credibility, as well as that of the findings, is preparation. The pre-audit dialogue, the conduct of the audit, and the informal real-time and formal final reporting are direct results of how well planning and preparation have been carried out. In Part 2, we will talk about this process. Knowing the operation and its hazards, the tools, equipment, materials and steps of the process, and some of the technology involved, lead to a meaningful beginning, and establish a common ground for discussion. These also help auditors plan the scope of their audits, where they will concentrate their efforts. We will look at and explain the concept of the maximum credible event (MCE), and how to target the truly significant hazards.

Part of the preparatory phase is to understand the requirements — what is legally, morally, or contractually required, and how compliance can be measured. These come from several sources. Understanding their applicability is essential so that the ensuing findings are cloaked with a mantle of legitimacy.

Safety auditing has been billed as a tool for excellence, as a management training methodology, and as a loss-prevention approach, among other things. It can only be these things when it is planned and executed for maximum effectiveness.

Chapter 5
Knowing the Hazards

To know the hazards is to know what needs to be done to mitigate them. No corrective action is nearly as persuasive to the doers as one that clearly addresses an acknowledged hazard, and one that has heretofore eluded a practical solution.

This is not to suggest that safety audits will always produce the long-sought answers to all of the unsolved problems that operating management has lived with day to day. It is to suggest that safety auditors should know what they are talking about when they identify safety issues and offer solutions to them.

The only way we know of to attain the credibility so crucial to plausible safety audits, and the ensuing acceptance of the specific findings made, is to know the operations under audit nearly as well as those managers whose operations are being audited. This challenge is met only through thorough preparation and study, from which safety auditors learn the technology of the operation, have determined the scope of the immediate audit, and have totally familiarized themselves with the operative paperwork.

Knowing the Operations

It is all well and good to suggest that any competent safety professional, armed with the OSHA regulations, can conduct a safety audit of any operating facility. He or she will undoubtedly do a creditable job, insofar as there will be recognition and reporting of discrepancies from the written rules. One doesn't have to understand the physical chemistry of a process, nor be competent at stress analysis, to determine whether egress paths are wide enough, if flammable storage cabinets are grounded, or if a pulley guard is in place.

However, that is not true auditing. If we wish to distinguish audits from inspections, then OSHA compliance verification is really the latter. OSHA compliance is a part of an audit, but only a part. A safety audit is intended to compare "as is" with "should be." What should be encompasses many facets in addition to OSHA compliance. Such areas as internal event reporting and analysis systems, management activity and involvement, operating sequences, housekeeping standards, equipment suitability, the adequate application of technology to operations, are only vaguely alluded to in the OSHA regulations. The conscientious training of workers, their conscientious adherence to safe work practices, and the effect of these on esprit de corps, safety, quality, and costs are unknown intangibles in the rule books. Furthermore, no OSHA, EPA, NFPA, TSCA, RCRA, or other regulation can state the adequacy, correctness, or technical appropriateness or reliability of processes, equipment, or procedures. There are rules on layouts, protection, egress, effluents, and communications, but not on engineering excellence. There are rules for handling accidents and injuries, but not on how, specifically, to prevent them.

The more that safety auditors understand the processes and technology of an operation, the more thoroughly they can identify the latent hazards, failures, or errors that can cause accidents. For example, inspecting a reactor kettle in a chemical plant might entail looking for the current calibration of the temperature and pressure gauges, seeing if there are leaks in the fittings, checking the housekeeping and for slick floors, whether there is a spill control and containment plan, and whether the needed personal protective equipment is available and in use. The safety inspector will probably ask for Material Safety Data Sheets and approved operating instructions.

Safety auditors who are knowledgeable about the operations will assess the *adequacy* of the personal protective equipment, the *correctness* of the operating sequence, the *adequacy* of emergency planning, etc. They can only do this if they understand the materials and equipment involved. They can anticipate that an erroneous sequence of materials addition, or a misread pressure gauge, or an air leak in a vacuum line can cause an aberration of the process, and possibly lead to a rupture. They can then evaluate the controls or redundancies that are or should be present to mitigate the hazards by detecting the causing events and controlling them.

Similarly, safety inspections of machining operations will determine whether machinists have their safety glasses on, whether drive mechanisms are adequately guarded, and that machinists use protective gloves for chip removal. Knowledgeable auditors will, in addition, be aware that safety can depend on the proper matchup of workpiece and tool materials, the proper selection of coolant material, and the correct selection of feed and speed parameters. They will either assess these independently, or raise the questions, so that they derive a positive feeling about the way in which the machinists or the shop planners, address such matters.

The point is, the auditors may not even address such questions, if they have no knowledge at all about machining operations and have only evaluated rule compliance. Safety auditors have to look past rules and regulations to safe production.

Different kinds of industrial activities have differing hazards, both in kind and degree. A safety audit in a textile mill will not be performed exactly like it would be done in a petrochemical or fertilizer plant. A refrigerator assembly line has very different hazards from an electroplating operation. Knowing the operation and its intricacies allows safety auditors to focus on the real hazards at hand, and to make sound recommendations to combat real hazards.

Three schools of thought exist in planning an audit. One is to use a topical or general checklist as a memory jogger, and to evaluate various items as appropriate. A comprehensive general checklist can get quite lengthy. We have put one together, which is shown in Part 3 of this book (on pages 175–183). Another approach is to tailor checklists to the operations, based on the knowledge and understanding that safety auditors bring to their audits. A third, and far less reliable method, is to use no checklist at all. We tend to lean toward the comprehensive, though general, checklist approach, because we rarely feel so omniscient that we are sure we have looked at all of the angles needed for tailoring a checklist. Neither do we consider our memories so infallible that we can be comfortable without any checklist at all.

Comprehensive and Specific Audits

We espouse and conduct two types of safety audits. *Comprehensive audits* are attempts to evaluate *all* aspects of safety, to discover *all* possible or suspected hazards. *Specific audits* focus on particular problem areas or hazard types.

For example, a comprehensive safety audit of a grain elevator operation in Nebraska might cover facets such as material handling; electrical codes; dust explosion control measures; access and egress; confined space control; fire detection, suppression, and protection; emergency response procedures and evacuation plans; OSHA logs and worker's compensation records; personal protective equipment availability, appropriateness, and usage; MSDS files and availability; housekeeping, and so on. A specific audit of that elevator operation's electrical system would be appropriate, for example, if there has been an unexplained series of circuit breaker overloads.

Specific hazard audits are often the logical reaction to identified problems. Comprehensive audits are proactive, preventive mechanisms for loss prevention and minimization. Both have their place in a total safety program. Even when reactive, specific audits should be viewed as preventive.

Specific audits need not be reactive, however. In an operation where one type of hazard has an obvious and overwhelming potential for loss, it is clearly a proactive and appropriate decision to conduct a wall-to-wall specific audit. In a refinery, for example, a fire detection, suppression, and containment audit is an obvious choice. In a paper mill, it might be machine guarding or static control. In an electroplating shop, it very well might be personal protective equipment usage and practices, and/or hazard communication program effectiveness.

Specific area audits have the advantage of focusing on the high-risk hazards in a relatively short time. They are based on historical information and present knowledge of what those major hazards are. Comprehensive audits have the advantage that they are designed to discover the "sleepers," the not-so-obvious hazards or deficiencies that can lie unnoticed, then jump up and bite. It is clearly a longer-term project compared to a specific audit of the same area.

Audit programs should, therefore, be planned with available resources defined and in mind. Time, manpower, and priorities are all parts of this equation. One final piece of advice—take chewable bites. In other words, the scope of a particular audit should be limited by geography, operational entity, or organizational element so that the scope can be comprehended, and progress toward completion seen and measured. Too expansive a scope, an overly ambitious audit, can overwhelm the safety auditor and delay delivery of important findings.

In a large plant, with dozens of operations and perhaps thousands of people, a plant safety audit is simply too huge an undertaking to contemplate. An audit of one of those operations at a time will give the safety auditor a scope that can be worked with and results achieved, and provide that operation's management with information in a timely manner.

Reviewing the Paperwork

Pre-audit paperwork reviews are advisable. They familiarize the auditors with the operations and organizations that are to be their subjects of study. It also allows them to gain that conversance without intruding upon the time and space of the auditees. Of course, they cannot in themselves provide total understanding or reveal all of the hazards, but they are recommended introductory exercises. A significant product of comprehensive paperwork reviews will be the formulation of knowledgeable questions and the identification of subjects that bear further probing or investigation.

Fundamental elements in paperwork reviews are the assessments of the policies, procedures, and systems that management has formally endorsed and espoused. These obviously give safety auditors a yardstick against which they can measure performance, attitudes, and compliance. More than that, they give auditors some insight into the adequacy of the safety system. If the policy statements are comprehensive and explicit, then what they find in the plants are a direct measure of how seriously the statements are taken, or how much latitude operating management has in implementing them. If the policies are superficial or vague, with no real system in place for implementation and monitoring, safety auditors will be forewarned of probable problems.

We do not recommend that management systems be evaluated in

these reviews. These conclusions should be formulated in the actual face-to-face discussions with operating managers and/or the safety professionals assigned to the operations. Again, the purpose of pre-audit reviews is to sense the culture of the organization, and develop topics for further probing.

At this stage, safety auditors should also have copies of operating instructions or procedures; local safety rules (for the operation or department being audited); accident records; training records; repair and maintenance records; and applicable local, state, and federal regulations under which the plant operates. These should be requested of the operating management at the first opportunity.

The benefits of reviewing the operating procedures, rules, and instructions are similar to those obtained from management policy reviews. They should give auditors information on materials involved, quantities used, equipment and machinery used, supplies and tools, the nature of the product(s). This basic understanding will help safety auditors anticipate categories of hazards. If toxic or flammable materials are identified in the operating planning, safety auditors will know, before they go to the floor, to focus on personal protective equipment, ventilation, or fire protection and control. If there is high-pressure or high-voltage equipment involved, they will have other obvious hazards focused. Quantities and throughputs, crewing, and shifts all help to define the nature and scope of the operation. Multiple shift operations give auditors a clue to look for shift turnover disciplines. The list of interactions goes on and on.

These reviews will also help auditors recognize which regulatory materials are applicable. They can flag sections of OSHA regulations and NFPA codes, perhaps identify ANSI standards of interest, or determine whether a military manual applies.

Another very useful paperwork review item is the training records. Auditors can get insight into the breadth and depth of training, or even whether a training program exists. Specialized skills and certifications also help auditors assess how valuable management considers training to be. If they find from the operating instructions that there is a considerable use of forklift trucks, but no record of who is trained to use them, a management system problem should be immediately suspected. Or, if the floor planning paper shows the use of significant quantities of, for instance, benzene, and the operating manager produces a list of all the people trained in

Hazards Communication, Respirator Fit and Use, and Handling Flammable Solvents, an auditor can have a pretty positive feeling that training is in fact a cornerstone of the safety program.

Repairs and maintenance records (or their lack) can tell auditors something about equipment failures, management attention and discipline in maintaining safe equipment, and may even distinguish band-aid repairs from corrective repairs. A good preventive maintenance discipline is a valuable clue to management's proactivity.

Drawings and layouts are extremely helpful in pre-audit reviews. Some special tool drawings or equipment installation drawings can give insight into how processes work. Plan and elevation drawings can identify potential congestion problems or egress deficiencies. Also, of course, the comparison of the drawings with the actual site can tell safety auditors something about how management maintains their original plan, or adapts from it.

To reiterate, paperwork reviews should be conducted as self-educational processes. A list of questions should be written down as they arise from the procedure, process, and policy records, and drawing reviews. Knowledgeable questions can be a positive beginning to a safety audit, which can otherwise be received as negative by the auditees. It will be rare indeed where the operating people acknowledge that safety auditors know as much about their operations as they do. It is almost a given that they will receive and view auditors as ignorant of their world. But the narrower the gap, the more credibility safety auditors bring to the opening contacts.

A fair question to consider at this point is what to do if the operating management, engineering departments, or process engineers indicate reticence or refusal to provide requested paperwork. Our advice in this case is to simply record that as a finding. Auditors should not threaten or accuse, simply document the facts. If that situation endures throughout the audit, it is a reportable finding to their superiors, their clients, or individuals that requested the audits. If the original refusal is reversed, it is probably wise to drop the issue and go on from there.

In summary, thorough paperwork reviews are invaluable preaudit exercises. They will make auditors vastly more plausible when they greet operating managers and walk onto the operating floor. Auditors who show up with no inkling as to what they are auditing are bound to become laughingstocks in the area. Auditors who speak the language and know what the major hazards are will immediately establish a level of credibility that will be extremely helpful through the duration of the audit.

If auditors know the nature of the operation, its products, its raw materials, its hazards, the kinds of equipment used in its manufacture, the management policies and procedures in place for governing the operation, and the necessary controls for coping with those hazards, then operating managers will likely interface with safety auditors as colleagues, rather than as adversaries. One prime objective of safety auditing is to be received as anything but an adversary.

Chapter 6

Prioritizing

Focusing on the truly important safety problems is one of the axioms of effective safety auditing. This is not to suggest that minor compliance and system deviations should be ignored. They must be recorded and reported very faithfully. But there are, inevitably, those findings that warrant special attention and flag-waving, because they are crucial to the safety of the operation.

This chapter is devoted to the determination of what hazards are truly critical, and how to handle them. Concepts such as what accidents can happen, how serious they can be, what predecessor events might cause them, how likely these serious accidents are, and, finally, how to zero in on the scenarios that have the most loss potential.

It is only after having studied an operation in enough detail to address these questions and issues, that safety auditors can focus their primary attention on the hazards, and their control, that could spell disaster to an operation. Again we say, this setting of priorities does not abrogate the auditors' responsibility to observe and report all of the comparatively minor errors and discrepancies that come to the fore during an audit. No one is so omniscient that he or she can forecast that minor deficiencies will only result in minor losses. The apparently minor deficiencies must be recorded also.

However, this chapter is designed to put safety auditors on the trail of the "biggies," and deal with them in an appropriate way.

What Can Go Wrong?

In the last chapter, we urged that safety auditors *know the hazards*, and study and understand the processes and technology, to learn what is processed, how and with what, by whom, and how much.

The point of all this planning and preparation, and getting acquainted with the specific operations, was to generate at least a partial appreciation of the nature of the operations, and a basis of commonality with the operating managers. However, an equally essential outgrowth of that study and ultimate understanding is the insight needed to address the title question above.

This insight, and the identification and listing of potential accidents, can be approached deductively in a tiered, or sequential, manner. To illustrate this concept, let us suppose that a paperwork review reveals that the operation involves a highly toxic and flammable material, such as methanol. The obvious things that can go wrong are fire and toxic poisoning. The next piece of information that an auditor might want to know is how much methanol is involved in the process operation. The paperwork review might indicate that it is only used in pint quantities to manually clean plastic moldings prior to an adhesive bonding step. The safety auditor might then surmise that the fire hazard is relatively small and localized, and that the toxicity hazard probably affects only the person doing the cleaning. The auditor might decide at that point that he or she needs to look for local exhaust, a nearby fire extinguisher, electrical code compliance, and the use of proper respiratory, face, and hand protection. The auditor might also add a note to verify that pint quantities are, in fact, the actual limit, and that spare bottles aren't stashed away in someone's tool box or drawer.

If, at the other extreme, the paperwork review indicates that a 500-gallon reactor, with methanol as the reaction medium, operates at 300 psig and 170°C, the whole picture changes. The fire and toxicity hazards take on entirely new meanings. Fireball and blast hazard analyses; cloud dispersion studies; stress and fatigue analyses; fire detection and suppression systems, and their testing and maintenance; emergency shutdown and backout procedures; remote control capability; pressure relief and venting systems; and redundant monitoring and control instrumentation would be some of the hazard control elements that the safety auditor would inquire into, to assess how well losses from those same two hazards are controlled.

A corollary question to how much is how many. How many people and what facilities and equipment are exposed? If twenty people are working side by side all day, and each is using a pint of methanol to clean parts, the exposure to both fire and toxic vapors is much greater than if one person, only occasionally, cleans one part.

Let's try one more example. Suppose a paperwork review of drawings and procedures indicates that the operation involves final contour machining of fiberglass moldings. The hazards that occur immediately are moving machinery, possible dust explosion, and very likely, a dust respiration and/or skin exposure hazard, perhaps both nuisance and toxic. The things one can envision going wrong might be exhaust ventilation failure, machine guarding errors, interlock failures or omissions, and failure to use proper personal protective equipment. *Total* loss control considerations would also explore the potential for machining configuration errors, leading to rejected parts, but we will limit our discussion here to safety problems. A safety auditor, having identified some of the things that can go wrong, is better prepared to look and ask about these potential problem areas. The auditor will look for, and perhaps ask for a test of, a ventilation failure alarm. Velocity measurement records can be requested, and a measurement obtained on the spot. He or she will surely look very closely at the physical condition of the machinery and its guards. The auditor will delve into the presence, absence, or bypassing of safety interlocks and sequencers. Also, he or she will obviously observe very closely the floor disciplines regarding personal protective equipment usage. Available data on dust concentration measurements can also be requested.

We do not mean to imply that this prior determination of what can go wrong should be the total scope of the audit. Keep in mind that this phase of the audit is Planning and Preparation, and that the deductions made here are part of doing the homework—getting up to speed on the operation and its hazards. When auditors get to the operating floor, it is almost inevitable that numerous other "can go wrong's" and questions about them will present themselves to safety auditors who keep their eyes open.

Another approach to determining, at least in this initial review, what can go wrong, is to look at actual and potential energies. This should be supplemental to the deductive approach we have discussed above, if used. Both real and potential energy are possible hazards if out of control. For example, elevated platforms are a fall hazard represented by the potential kinetic energy of mass times height. Moving machinery is a potential hazard in terms of actual kinetic energy ($wv^2/2g$). Reactive or explosive chemicals and mixtures are potential thermal and pressure hazards, in addition to the possibility that they may present a health hazard. Electrical equipment poses an obvious energy source that can, and often is, hazardous. Pressurized vessels, pipes, and fittings have stored energy (potential) that can become actual kinetic energy in a rupture, besides the possible health hazard from the resultant toxic release. The list can go on and on.

A pre-audit assessment of the forms and magnitudes of both potential and actual energies can lead the safety auditor directly to some of the things that can go wrong. Any source or form of energy that is uncontrolled has a consequent result that is probably "wrong."

Finally, we suggest, during this pre-audit assessment, that the past history of accidents and injuries, and any previous safety inspection and hazard analysis reports be reviewed in some detail. Obviously, the history of accidents and injuries is the story of what has already gone wrong. How well accidents were documented and investigated, and how complete and effective the corrective measures were, are items that the safety auditor will assess later. However, given that the accident and injury record discloses exposures and hazards, the auditor has some immediate clues as to where to focus his or her attention, while formulating an overall plan. The previous inspection and analysis reports are additional clues, in a sense, to the auditor who is retracing someone else's steps, to the extent that they clearly document what someone else thought could go wrong. We want to emphasize that it is not essential that safety auditors deduce and validate scenarios showing the exact mechanism by which these events can occur—how the things go wrong. In our estimation, the mere presence of methanol, or the simple fact of moving machinery, should be a signal to auditors that a hazard, a mechanism for an accident or injury, might exist. It alerts them to the need for further evaluation when they reach the operating floor. They do not have to—probably should not—discuss their lists of things that can go wrong at this stage. Making up such lists simply gives them some direction when they start their floor audits.

The Maximum Credible Event (MCE)

An understanding of hazards and their possible consequences affords an intelligent basis upon which to develop audit plans. A thorough deduction of "what can go wrong?" lists give safety auditors direction and purpose when they go out on the operating floor. Never mind that the list is not 100 percent complete or accurate. It simply forms a backdrop against which observations and inquiries can begin.

This listing of potential problems will inevitably include a fairly wide variety of items, with an attendant variety of consequences, ranging from the trivial to the serious, and possibly even catastrophic. In such a listing, there will be relatively few items that will conjure up consequences that are blatantly unacceptable, too injurious, damaging, or threatening to tolerate. This is the maximum credible event, the MCE. While we talk about *the* MCE, the fact is, that in any given operation, there might be several events whose consequences qualify them as MCEs. It is really of little benefit to attempt to deduce which of a number of totally unacceptable potential accidents is *the* MCE. All MCEs must be controlled or contained to the best of our collective abilities.

So what constitutes MCEs, and why do we want to identify them? Let's start with the first question. We suggest, first of all, that the key operative word here is "credible." Not only the event and its consequences must be credible, but so must the circumstances, the precursor conditions under which it can happen, be credible. Past serious or disastrous accidents carry a kind of automatic credibility, obviously because they actually happened. The control and containment measures developed to cope with such events tend to become integrated parts of the industrial culture. Rare indeed is the new commercial structure that is neither fireproof nor sprinklered, because the MCE—destruction by fire—of flammable unprotected buildings is real, proven, and wholly credible.

Other MCEs are intuitively obvious, recognized by one and all, and may or may not be totally controlled or contained. Every time an airplane rolls down the runway, the obvious, universally recognized, and accepted MCE is that it can crash. This is a case of a partially controlled MCE, which the airlines, passengers, and society as a whole have come to accept as a way of life. Safety auditors will occasionally encounter these as well. We hesitate to try to cite an example of a recognized and totally controlled MCE. We are not sure there is such a thing; it seems like the concept is similar to that of the universal solvent. If it is totally controlled, then is it credible?

More difficult to pinpoint, much less prove, are the envisioned, hypothesized MCEs, the ones deduced by reasoning. Here is when credibility becomes essential. To be credible, or plausible, or believable, the event must be one that an informed person can agree *can* happen. There must be a triggering agency in the operating environment that all agree is really there. The conditions, failures, or mistakes that can pull that trigger must be ones that knowledgeable people can agree *could* occur—they are possible. The chain of such errors and failures may be highly sequence-dependent, or they may be totally random, but they must be credible! Finally, the deduced consequences must be shown to follow from the event in a credible, logical sequence. Further, whatever external factors that might be necessary for the consequences to reach their maximum conditions must be identified and shown to be credible.

It is probably time to illustrate these possibly nebulous concepts with a couple of concrete examples, to try to show the thought processes that might be followed to zero in on the MCEs and their state of control or acceptability.

Consider a horizontal autoclave operation that involves thermal treatment of large laminated structural members under a high-pressure atmosphere of carbon dioxide (CO_2). Typical operating conditions might be 250 psig and 150°C. The operating sequence might begin with a partial evacuation of the ambient atmosphere, followed by introduction of the CO_2, gradually increasing both pressure and temperature until the target conditions are reached. These parameters are then probably held for some predetermined time interval, perhaps an hour or two. This is likely followed by a cooling cycle and a blowdown of the CO_2. Finally, there is probably an automatic air purge of the autoclave to make the atmosphere inside breathable. In many modern installations, the entire sequence of door closing, latching, CO_2 valve and heater activation, pressure and temperature monitoring and control, timing, coolant flow, and air purge actuation and unlatching will be sequenced by a programmable controller.

A very neat and tidy operation! The operators load the parts into the autoclave, keep an eye on the control panel, and take the parts back out. But we would immediately identify at least two MCEs. First, the autoclave, being a pressure vessel, has some risk of rupturing, due either to overpressure, or to some undetected deterioration of its structural integrity. Second, a person remaining inside this confined space when the door closes, and the controller is activated, would be an obvious fatality.

The safety auditor is immediately alerted to quite a number of items to look for and verify when inspecting that autoclave. The auditor will look for redundant pressure relief ports, testing records, design calculations to balance relief capacity with volume, redundant pressure and temperature sensing capability, maintenance and inspection records, etc., in order to assess how thoroughly and effectively the rupture MCE has been controlled. Note that none of the control features has eliminated the MCE. The only way this MCE is eliminated is to render the autoclave incapable of pressurization.

For the trapped person MCE, such features as voice-actuated emergency lights, internally actuated alarms, manual internal unlatching and controller overrides, and training records for how to use them, can be checked out by the safety auditor. Again, the trapped person MCE has not gone away, even with all of the precautions mentioned—or others. That is the irony and the constancy of the MCE concept. Nothing short of going out of business will make them disappear. This may sound frustrating, even hopeless at first. But it's usefulness is realized once that initial mental block is passed. The purpose of identifying MCEs is *not* to eliminate them, but to come as close to controlling and containing them as our resources and ingenuity allow. They say—"this can happen—are you satisfied with the probabilities?"

Will another example help? At the risk of driving the topic into the ground, we will offer one. Robotics is a current technological frontier; sophistication and applications in this field are growing rapidly. The safety auditor is quite likely to encounter them in an operating area sooner or later, especially in manufacturing and assembly operations.

A typical floor- or wall-mounted robot has an MCE that everyone conversant with them is aware of, and has long since acknowledged.

An out-of-control robot can maim. There are even a few reported deaths resulting from a robot arm pinning a victim against something. Numerous precautions, interlocks, interrupts, electric eyes at the perimeter of the "reach" envelope, training and warnings, etc. have rendered the use of robots "safe," as well as efficient, productive, and economical. But their ability to cause injury remains. The MCE is intact after all those precautionary measures have been taken. The degree of hazard control, however, has reached an acceptable level and robots have gained their place in industry.

One other point needs to be made about MCEs. Some of the most serious ones are not necessarily associated so much with the operation that we are about to audit, but with the abnormal situation, the abberation or upset in the operation, that prompts on-line repairs, trouble-shooting, etc. This is a situation in which line operators may be tempted to take short-cuts, expose themselves to a hazard "for just a second," to fix the problem. These hidden hazards, these unplanned and unforeseen actions, are much more difficult to envision in this planning and preparation stage of the audit. Our advice is to simply recognize that at this point there is still much to be learned when the floor observations begin.

This does raise the question of maintainability, however. One of the things that safety auditors should be aware of when they are on the floor is the accessibility and availability of the equipment that might be prone to failure or in need of adjustment or repair. It is often the case that installation designs ignore the very basic element of accessibility, in favor of operational elegance. The result can be built-in entrapment of the operating people who try to make the equipment run as it was intended.

What Can Cause an MCE?

We touched on the need for safety auditors to not only identify MCEs, but to establish their credibility. Otherwise, by definition, they are not MCEs. The credibility of MCE causes must be established hand-in-hand with the events themselves. Falling off of the roof of a building is not an MCE unless there is some way to get to the roof. Once a credible way is found to get onto the roof, then the fall is established as an MCE. The credibility of the cause of, or sequence of events leading to, the MCE is what establishes the MCE.

We do not wish to get entangled in semantics, but we are well-advised here to make some definitions. Any accident has some number of precursor acts, events, or conditions; occurrences that have the potential to damage or injure; and a result or consequence. An MCE is the worst plausible result of an occurrence. In the simple example above, precursors might be climbing a ladder to get onto the roof, not tying off to the structure, and misjudging footing. The occurrence is the fall. The consequence might be anything from bruises to broken bones to death, and certainly the MCE, for which the auditor will seek control measures, is the latter.

So it follows that the perceived consequence of the occurrence must have credibility, along with the precursor events and the occurrence itself. But we are dealing in this section with causes, on the premise that the MCE has been deduced to be rational and credible. Also, we want to distinguish here between proximate causes and systemic causes. Maybe a more realistic example will illustrate our point. Punch presses have a nasty habit of punching people's hands. This is a bona fide MCE. But how and why does it happen? What are the precursor events, the causes of the MCE? For a punch press to injure its operator, the operator must have had his hand in the action area of the press. Also, the safety features designed into the installation to preclude actuation when hands are present must have been defeated. This might occur because the two-handed control buttons were mischievously overridden, or because of a failure in the circuitry that controls the press actuation. In either case, the immediate, proximate cause is intrusion of the hands into the action area of the press when it cycled. This has nothing to do with the schedule-driven systemic causes that may have coaxed the press operator to override the controls, or the foreman to waive needed preventive maintenance inspections.

Let's try a more dramatic example. An operation that we are familiar with is the drying of a wet, powdered explosive. The MCE for virtually any explosives processing operation is the detonation of that explosive, when operating people are present. The second-level MCE is to have a detonation when operating personnel are not present, which is the idea behind remotely controlling such operations. After charging the wet explosive to the dryer, the operators retire to a control bunker that affords protection from the blast effects of the MCE. During the course of a seemingly routine drying cycle, the dryer, and the building in which it is housed, are scattered over twenty acres of real estate. The second-level MCE occurred. Why?

There are any number of technical, mechanical reasons why the dryer detonated. We discovered two probable reasons in the course of the investigation. It could have been improper maintenance or improper design. These are both engineering causes.

The logical question now is, could a safety audit have had any hope of averting the explosion? Our answer is an unqualified maybe. A safety auditor who understood the nature of the explosive being processed, and who had reviewed the paperwork in detail (in this case, the design specifications and drawings, and the preventive maintenance checklist), might very well have determined that the failure of internal securing mechanisms, or the fatigue of critical welds, might occur, and that these failures could result in the MCE. Whether the auditor did deduce this or not, he or she should certainly have seen that the MCE is an explosion of the dryer, and that all possible means must be used to ensure that it be operated remotely.

Identifying MCEs is central to an effective safety audit. Identifying the causes of the MCEs is something like icing on the cake. It gives safety auditors focus for their investigations and observations. The danger in identifying causes is in being lulled into a mind-set that only those identified causes are valid. Causes are leads, and need to be followed, but safety auditors must be fully aware that they are not so omniscient that other causes might not become apparent in the course of their audit.

Likelihood and Seriousness

MCEs can have any and all degrees of probability. How likely the MCE is can have a major influence on how energetically its control is pursued. Establishing credibility makes the MCE viable; determining its likelihood and seriousness makes it a force to be dealt with.

Most safety professionals will be acquainted with the hazard matrix, where the likelihood and severity of an accident are evaluated as objectively as possible, with the result being used to categorize the hazard level. Typically, the severity is classified as catastrophic, serious, moderate, minor, or negligible, while the likelihood may be graded as certain, probable, improbable, or impossible. Thus, a catastrophic accident that is certain to happen under some specified set of circumstances becomes a Hazard Level I, and positive action is imperative. Similarly, an improbable event that can only result in negligible damage or injury will be a Hazard Level IV. Intermediate probabilities and severities will yield intermediate hazard levels. Table 6.1 details one such matrix we are familiar with.

We bring this up because this hazard classification methodology is a very valid approach to dealing with MCEs. The probability and seriousness of the accident will have a major influence on the rapidity and thoroughness with which corrective or containment actions are prosecuted. An MCE that is catastrophic but impossible is not an accident that must be prevented. It is not a true MCE. But one that is catastrophic, though improbable, must be dealt with to make sure that the conditions that make it improbable are valid and remain in place. Furthermore, deducing the likelihood and seriousness of an MCE, and reaching a conclusion on the hazard matrix level, automatically alerts the safety and operating people to the type of control/containment measures that are appropriate (see Table 6.1).

There are, of course, other ways to assess the likelihood and seriousness of potential events. As we have mentioned earlier in this work, past history is very valuable. The very nature of a previous accident tells us a lot about both its probability and its severity. The fact of its occurrence and how damaging or injurious it was are in the record, providing reasonably careful recordkeeping has been practiced in the past. Also, if such records are available, safety auditors can hopefully determine whether conditions existing then and now can result in the same MCE, whether the likelihood has been reduced, and whether the severity has been in any way effected.

Analytical approaches can be useful in assessing these aspects of MCEs, also. Previous formal hazards analyses, or specific technical studies, are very strong clues to the prior assessment of a perceived accident's credibility and result. Lacking these, safety auditors can do their own informal, perhaps amateur analysis. An intuitive evaluation of probability and severity, based on a knowledge of the process and its hazards, is not as unprofessional as one might suppose. The whole reason for making such an evaluation is to provide

Table 6.1. Hazard Level Matrix.

		A	B	C	D	E
				Probability Level		
Severity Level	1	I	I	II	III	III
	2	I	II	III	III	IV
	3	II	III	III	IV	IV
	4	III	IV	IV	V	V

Definitions:

Probability Levels:
    A Certainty (or close to it)
    B Very Likely
    C Likely Eventually
    D Improbable
    E Virtually Impossible

Severity Levels:
    1 Catastrophic
    2 Serious damage or injury
    3 Minor damage or injury
    4 Not hazardous

Matrix Levels:
    I Imminent danger—Immediate engineering corrective action required.
    II Serious danger—Engineering or safety devices required.
    III Moderate danger—Engineering, safety devices, or procedural controls required.
    IV Minor danger—Safety devices or procedural controls required.
    V Negligible danger—Corrective action optional.

an initial focus for the audit, to zero in on the hazards that cannot be allowed to go uncontrolled. We have found the matrix approach in Table 6.1 to be as exact as is necessary for this purpose, even though it is admittedly a wholly subjective method.

Given only five degrees of likelihood for an MCE, two of which are "certain" and "impossible," it does not seem too taxing to intelligently, even though subjectively, determine which level of probability fits the situation. Similarly, four levels of severity can be rather easily distinguished if safety auditors have done their homework in understanding the process and hazards. When this exercise tends to get bogged down, due to an honest inability to grasp the technical nuances of an operation, we strongly recommend consultation with engineering and operating people, rather than making a hasty and erroneous deduction. A consensus of knowledgeable people is more likely to be close to reality.

Targeting the Right Hazards

Getting a good understanding of the MCE, its potential causes, its probability of occurrence and the severity of the consequences if it does occur, will lead safety auditors to an assessment of *the right hazards*. We do not want to imply that only abatement of MCEs should be evaluated. The objective is to make sure that they are not forgotten in favor of more tangible visible problems. If we err through omission, we don't want it to be with the serious problems.

Table 6.1 in the previous section gave a very brief definition of hazard levels and the corrective/control/containment measures appropriate to each. We want to elaborate on those ideas here. A hazard level is deduced from a subjective understanding of the likelihood and seriousness of a credible accident. The higher the matrix-derived level, the less ultimate concern we need have about the accident. It also follows that the control measures need be less and less rigorous. The obvious way to get the maximum amount of proactive accident prevention for a dollar is to deal with the lowest hazard matrix level events first—not *only*, just first.

A Hazard Level I needs immediate attention. It is essential that safety professionals *immediately* notify operating managers of their conclusions, however tentative, and request a review of their methodology in making those deductions. Safety professionals should, as succinctly as they can, explain what rationales were used to arrive at these fears, and look for measures in place to counter them. If the operating managers are not conversant with the concerns, or cannot knowledgeably describe what measures are in place for handling them, the safety auditors should request, or at least strongly urge, that the operations be shut down until the perceived MCEs can be more thoroughly evaluated. We recognize that there is a "red herring" potential in this, but rationally derived Level I MCEs are not to be dismissed lightly.

If the Auditors' rationales were sound, hopefully someone else once perceived the same MCE and took positive and permanent actions to abate the level to a higher (less severe) category. The safety auditors then can evaluate those abatement measures to satisfy themselves that they are in place and effective. A Level I hazard should never depend on safety warnings or procedural discipline alone to cope with it. It is, by definition, likely to happen and to have severe consequences. Management should not depend on bells, whistles, or operating procedures to mitigate such dangerous conditions.

A Matrix Level II hazard is defined as serious. It can take several forms of logic. It may be something that is certain to happen, perhaps often, but has minor, non-life-threatening consequences attached to it. A Level II can also, however, represent a hazard that might occur fairly infrequently and have very serious results. It might be likened to the Russian roulette of hazards levels. Only fools will let them go uncontrolled.

Level III hazards range from events that are virtually certain to occur, but would be nonhazardous, though those that are likely to occur, but would be of a minor nature, to events that are improbable or impossible, but would be serious if they did occur. Taken as a class, they are called "moderate" hazards. They must be controlled, but not necessarily mechanically or electronically. Procedural controls are deemed permissible because they are deduced to be of little danger or because they can hardly be rationally expected to happen in the first place—they are imaginable, but barely. Any event that is both unlikely and of little or no consequence if it does occur reaches the Levels IV and V. In both of these latter cases, procedural hazard control is considered adequate or optional, respectively.

At this point, we would probably be well advised to illustrate these seemingly subjective deductions with a real-world example. First, however, we want to emphasize a point made previously when we introduced the concept of the maximum credible event—the MCE. The point is that the MCE does change or reduce when precautions are heaped on top of precautions. The hazard matrix level can be raised from high hazard Level I or II to low hazard Level IV or V with multiple and redundant control measures, *but*—the MCE remains the same. Its likelihood has been effected, not its result.

Let's use a large vapor degreaser for our example. Typically, such an installation consists of a tank, accessible at the top for inserting and retrieving parts for cleaning. In the bowels of the tank is a reservoir of cleaning solvent, and some method of directly heating that solvent. Usually either steam coils or electrical heating devices will be used to heat by submersion in the solvent reservoir. Near the top of the tank will be cooling coils to condense the solvent vapors created when the heating unit boils the solvent. When a dirty part of some kind is suspended inside the tank, the relatively cold part also acts as a condenser. The condensing solvent cleans the dirty part on contact, and drips off, carrying the dirt, grease (hence the term degreaser), oil, or grime with it.

Normally, the solvents used in vapor degreasers are very hazardous to health, even in small doses. Perhaps methyl chloroform is the most common, with a TLV (threshold limit value) of 350 ppm. When methyl chloroform is boiled, however, the concentration of vapor in the degreaser tank approaches 100 percent, and displaces the air as the vapor rises in the tank toward the condenser coils. Therefore, in normal, intentional operating circumstances, the tank contains a lethal atmosphere, virtually pure methyl chloroform. Is it clear by now that an MCE for this installation is death by suffocation or toxic exposure?

How can this MCE happen? What can cause it? Obviously, it can occur if someone goes into the tank. That person can fall in, jump in, or lower a ladder or rope and climb in. The MCE can also occur if the tank breaks, leaks, overflows, or boils over. Any way in which people can be subjected to an overdose of methyl chloroform vapors is a way in which the MCE can happen. Having established that the MCE is catastrophic, and that there a number of imaginable ways for it to occur, we get to the crux of this section—what is the actual hazard level?

Suppose we find this degreaser tank installed below ground, with the top of the tank almost flush with the floor. There are no barricades around the edge, and people are working all around the area. When they want to clean a part, they go over and hang it on a hook and lower it into the tank. Also, just to make this as unreal as possible, without losing imaginability, some of those parts weigh 80 pounds. Would it seem correct to surmise that the probability of someone falling into the tank is at least "Very Likely" by the definitions of Table 6.1? From the matrix of Table 6.1, we see that a Catastrophic Severity Level 1 and a Very Likely Probability Level B is a Hazard Level I—imminent danger, requiring immediate engineering correction.

So we engineer out the fall potential by putting guard rails around the perimeter of the tank and by providing an overhead hoist to lower and retrieve these 80 pound parts. The MCE is, of course, unchanged, as we have stressed before, since someone can now climb over the guard rail to free a tangled hoist cable, and still fall into the lethal tank. What is the Probability Level though? Is it less likely that operators will intentionally do that than that they would unintentionally fall while doing something else? It should be intuitively obvious that the answer to the second question is "Yes." The hoist is not likely to get tangled every time, and by virtue of climbing over the rail, operators are going to be very conscious of the gaping tank beneath them. Does it seem reasonable to give this situation a Probability Level C—"Likely Eventually"? That would be our assessment.

A hazard that is still Severity Level I, and likely eventually, is now a Matrix Level II—serious danger, requiring *more* engineering controls, or safety devices (warnings). So what can we do to make the asphyxiation or toxic exposure event Probability Level D—"Improbable"?, or should we accept the present level of risk? That latter question is properly referred to upper management for a decision, but safety professionals should strongly recommend against accepting the risk at Level II. We can still improve our chances, even against an obviously catastrophic MCE. We can put guides and slots to obviate entanglement of the hoist cable. We can provide a movable cover for the tank, or a catwalk with only handholes. We can virtually eliminate the possibility that someone can fall, jump, or get into the tank until it is stone cold and empty.

However, as we showed at the beginning of this example, spills, leaks, boil-overs, and overflows can result in at least a toxic, if not asphyxiating, exposure of people in the surrounding work areas. If we are to make all of these avenues to the MCE equally "Improbable," we will have to engineer interlocks, detection and control logic, and hardware that will prevent over-filling, overheating, detect escaping vapors, and sound evacuation alarms when a breach is detected. We will have engineered emergency cooling and quenching capability, ventilation systems with emergency backup power, and interlocked the heating system to the cooling system to prevent boil-over. This is not intended to be a lesson in vapor degreaser design technology, but (as a matter of fact) many modern degreaser installations have incorporated most, if not all, of the features we have rather blithely deduced here, plus many others.

In spite of all of those precautions, the MCE has never changed. We have merely reduced its likelihood to an acceptable level.

Chapter 7

Knowing the Requirements

Safety audits garner the maximum credibility when the audit findings are legitimate deviations from an acknowledged, authoritative standard or requirement. The legitimacy is only enhanced when the findings are compared with one of these sources. Few indeed are the operating managers who will flaunt a known requirement, which no one questions.

Few indeed also, are the operating managers who will not be touched by the reporting of compliances with those legitimate, authoritative standards. We might observe, once again, that the surest way toward total compliance is to play up the positive findings, where compliance is rampant, and treat the noncompliances in as matter-of-fact a way as the situation warrants.

Requirements are usually quite objective—there is often little room for negotiation as to what the requirement is, or whether it applies. We will discuss this aspect, along with the more controversial facet where professional opinion and analysis constitute the basis for a finding, a safety concern. We will also explore the general duty clause of OSHA, and consider some of its ramifications.

Hazards, Opinions, and Requirements

Hazards exist. Their magnitudes, and the probabilities of causing trauma, vary all over the spectrum, but we see them lurking in every area of our lives. The hazard, as we have pointed out, is generally in the form of an energy, actual or potential, or a material with an actual or potential ability to damage and/or injure. The existence of hazards is a *fact*. The fact of a hazard in the workplace should not even be a point of contention. Virtually every operating entity uses some form of electrical power, either to provide light, or to power equipment, or to heat or cool the environment. If there is electrical power in that operating area, then it should be taken as an irrefutable fact that electrical hazards exist in that area. What often becomes a point of debate or even argument, is *how* hazardous it might be. How likely is the acknowledged hazard to cause a loss? How serious might that loss be?

Chapter 6 treated these questions in a way designed to assign levels of hazard, and discussed what safety professionals should do with their deductions. Here we want to distinguish hazard identification from hazard containment or correction, as they are characterized by either professional opinion or rules and regulations of some kind.

Hazards do indeed exist, and the first step for safety auditors is to identify them and characterize their potential for loss. It is a fact of life that there are many hazards in the workplace for which there are either no governing rules or regulations, or those that may pertain are vague and subject to forced interpretation. In other words, there may not be a "requirement," a "should be" against which safety auditors can contrast "what is." Yet the hazards are there, and the auditors have concluded that the hazards are not safely controlled or contained.

They must clearly outline their rationales for these conclusions that the hazard levels existing demand mitigation to a higher level of control. But they will, in this situation, be offering their professional opinions, and we have already touched on the tenuousness of those grounds. If they have successfully shown the reality of the hazard to their potential challengers, their rationale will be the cornerstone of their recommendations for improving the situation. Any specific recommendations as to how to effect these improvements will be their own opinion. They must realize that others will have differing opinions regarding the likelihood of loss, the magnitude of loss, whether it warrants improvement, and what that improvement should be.

Opinions as to how an operation should be conducted should not be offered by safety professionals in the absence of a truly definable hazard, or a clearly stated requirement. For example, there is bound to be an instance, in some operation, where dirty floors are not a hazard in any rational evaluation of the operation. If, in addition, there is no management policy or directive requiring clean floors, the safety auditor is probably well advised not to give his "opinion" that clean floors must be maintained. A better choice, if the auditor has a strong opinion on the matter, might be to propose to management that a clean floor policy be adopted in the interest of improving the image of the area, improving morale and pride in the workplace, improving potential customer's impressions, etc.

Another aspect of professional opinions is that they should never be contrary to existing requirements. One of the surest and fastest ways for safety auditors to lose their credibility is to recommend one thing, and have a nonprofessional point out that it is in violation of OSHA or company policy or the local city ordinances. It is, on the other hand, often appropriate for safety professionals to offer the opinion that the existing rules and regulations do not go far enough in mitigating a hazard, and that extra steps are needed to adequately control it. In this situation, safety professionals are in much the same situation they were when there was a hazard with no governing requirements. They must carefully communicate their rationales as to why the hazard is greater than should be tolerated, why mere compliance leaves an unacceptable risk, and therefore why additional control/containment measures are called for.

Finally, if safety auditors are going to reach into their professional bags and produce opinions as to how to correct hazardous conditions, they must ensure to the best of their abilities that such measures are technically sound. If they have specific recommendations to make, a wise choice, in the name of maintaining credibility, is to seek credible verification from a recognized source. Specific recommendations that are technically impossible or foolish will erode credibility instantly. Safety professionals are better advised to omit any specific recommendations if they cannot obtain credible independent endorsement of their approach to corrective action. They should concentrate on communicating the seriousness of the hazard, and the need for experts to determine the best way to abate it.

Requirements that are clearly and unambiguously stated are the safety auditors' best allies. They can have the unique attribute of removing opinion and argument from the hazard identification and abatement sequence. There are a few thoughts that safety auditors should keep in mind when citing a written requirement as a reason for action, however.

The auditor should learn at the outset to present a requirement as a separate authority—*not* theirs. They must try never to give the message that this is their requirement. Let the written word speak for itself. If they assume proprietorship for the requirement, they will be back to the relatively weaker position of offering opinions.

Secondly, the requirement should clearly relate to the hazard being considered. The context of the requirement must be the same as the operational situation. Twisting a requirement to make a point is counterproductive. For example, when the requirement for guard rails on drops of 48 inches is applied to a truck dock, the misapplication of the requirement is obvious.

Third, it is almost axiomatic that an authoritative requirement overrides an opinion. Safety auditors can stand their ground firmly and politely with that kind of back-up.

We have tried in this section to provide some direction and distinction in the matter of hazard identification and control. Hazards exist by virtue of doing something with something. Whether they are adequately controlled boils down to either a question of compliance with a requirement, or a professional opinion as to whether they are. Both considerations have a valid place in safety auditing, but require different approaches to communicating with, and selling to, operating management.

Legitimizing Requirements

We have pointed out in the preceding section that opinions must be technically sound and that requirements must be valid and applicable. In other words, they must be legitimate. The cited policy or regulation must clearly relate to the situation at hand.

This raises the very obvious question of what constitutes legitimacy. It might seem that there is an equally obvious answer, but as we develop this concept, it will be seen that legitimacy is a complicated matter. Perhaps the most obviously legitimate body of regulatory material is the Occupational Safety and Health Administration (OSHA) Rules and Regulations (29 CFR 1910). Of course, in states that have opted to administer OSHA locally, there are equivalent state OSHA rules and regulations. If you read the introductory paragraphs of 29 CFR 1900, they leave little doubt about the legitimacy of the volume. On page v it reads, "The contents of the Federal Register are required to be judicially noticed (44 U.S.C.

1507). The Code of Federal Regulations is prima facie evidence of the original documents (44 U.S.C. 1510)."

We have no pretenses regarding our legal expertise, but this language would appear to be rather iron-clad to us. A requirement cited from 29 CFR 1910 has the force of law, as any practicing safety professional will attest. The only way a cited reference will lose its legitimacy would be to use it in an inappropriate circumstance. It is obviously bad form to cite a requirement for helicopter maintenance against a conveyor belt installation or a forklift.

There are other specialized regulations and codes that have a legal legitimacy similar to OSHA's. Title 10 CFR comprises the rules and regulations of the Nuclear Regulatory Commission. These have every bit as much legal status as 29 CFR 1910, and perhaps more. The Environmental Protection Agency and others have similar legislative origins and codified requirements with the force of law.

Once this basic fact is realized, it is rare indeed that an individual will take issue with the legitimacy of a requirement cited from such a source. The safety auditor's main task in this effort will probably be to educate (if necessary) the operating manager to the facts of life regarding federal or state regulations. The other, primary responsibility is to ensure that the cited reference is legitimately applied to the finding being reported.

A second kind of "requirement" includes the so-called advisory standards. These are such consensus standards as ANSI (American National Standards Institute) and NFPA (National Fire Protection Association). These "regulations" are extremely detailed and cover a wide range of subjects. Their legitimacy arises from two potential sources. In some cases, certain of their standards are incorporated into OSHA or state-OSHA rules and regulations. When that is done, they are, of course, no longer advisory. The other fact that lends legitimacy to such standards is their consensus status. They are not one man's opinion, nor even a small self-appointed committee's. They are the collective effort of experts in the various fields treated. While there may not be a legal need to comply with an advisory standard, there is a moral obligation to consider the advice contained therein, and to make a conscious decision to either comply, or develop a rationale for not complying.

The third class of requirements we want to mention are the in-house standards—the company's own written policies, directives, or procedures. To the savvy operating manager, these documents ought to carry as much clout as OSHA, if not more. This is providing that the top management considers their safety policies as something more than window dressing. If the general manager, or the president, or operations vice president has put his name on a safety requirement, it should have the "force of law" within that organization. (If it doesn't, that is a finding, too.)

A feature of a company directive that is different from an OSHA or NFPA standard is that they can be changed within the company. We have had many experiences where an obsolete or even wrong rule was on the books, we cited a deficiency against it, and the corrective action was a revision to, or deletion of, the rule. We were perfectly happy with that outcome, because the rule was clearly inappropriate. A caution here, of course, is that in-house rule changes must never override or negate a higher authority. The fact that a company officer can establish policy with a sweep of his hand does not give him the right to repeal OSHA.

Another valuable way to legitimize findings is through the recognized expert. This is somewhat similar to using a consensus standard, but much less anonymous. Someone renowned in a particular technology, material, or process, who can validate the safety auditor's finding, or suspected hazard, will lend a legitimacy to that finding, and may provide the solution at the same time. Consultants within an industry or technology can be a very valuable resource for increasing the legitimacy, and therefore the appropriateness, of the resultant reaction to safety hazards.

Suppliers of materials, equipment, and machines are another source of legitimacy. Manufacturers of equipment generally provide owner's manuals, operating procedures, maintenance and repair manuals, *and* safety warnings and procedures. What can more readily validate a safety finding than a warning from the maker of the machine? Similarly, suppliers of chemicals, compounds, and other formulated materials will, in fact must, provide Material Safety Data Sheets (MSDSs) when asked. They are required under OSHA law to prepare these hazard warning and control documents, and provide them to their customers and users of their products. Again, what could be more legitimate than safety requirements provided by the manufacturers of the materials themselves?

Somewhat similar to the advisory standards and the consulting experts are textbooks, periodical articles, courses, and training materials provided by a variety of public and private organizations. Materials available from the National Safety Council, its divisions and groups, and from the American Society of Safety Engineers and its divisions can be of considerable help in legitimizing concerns about a finding. They provide the expertise of a name consultant with the prestige of a national organization backing them.

Finally, the company's own accident history, and that of their industry, can legitimize a perceived hazard. Many industry organizations such, as the American Petroleum Institute, the American Institute of Chemical Engineers, the American Safety Society of Mechanical Engineers, and others sponsor safety seminars on topics of peculiar interest to that industry. Both the company's accident history and the industry's concerns can be valuable confirmations of what can be reasonably expected to go awry, thus legitimizing a finding of concern to safety auditors.

The General Duty Clause

There are two previously alluded-to ideas that form the backdrop against which this topic is discussed. We talked about the not-unusual circumstance where an intuitively obvious hazard exists and for which we can neither find nor cite a governing regulation that will remove the needed corrective action from the realm of the optional. Also, in Part I we discussed at some length the various styles and systems of management and operating managers that safety auditors can expect to encounter in their auditing activities. When auditors discover that they are dealing with a reticent manager about a genuine safety hazard with no specific written regulatory road map to control it, the auditors may feel that they have run into the proverbial brick wall.

They cannot conscientiously give up and allow the hazard to go unabated. However, they cannot hold out a company or legal document to, in effect, intimidate the obstinate manager into abatement actions. What other options do they have? We are assuming, in this scenario, that the operating management are the prototypical passive kind, who do not want to be bothered in the first place, and will expend no effort or resources they are not forced to.

This apparent impasse brings us to The General Duty Clause.

Part 1903.1 of 29 CFR 1900 states, in part, "The Williams-Steiger Occupational Safety and Health Act of 1970 . . . requires, in part, that every employer covered under the Act furnish to his employees employment and a place of employment which are free from recognized hazards that are causing or are likely to cause death or serious physical harm to his employees." This Part goes on to require compliance with OSHA standards, rules, and regulations among other powers vested in OSHA by the United States Congress. But the subject of this section is the quotation above, endearingly known the nation over as the general duty clause (sometimes called the GD clause).

The general duty clause is perhaps the most famous verbiage in the OSHA Rules and Regulations. One reason for this fame is its often-used basis for employer citations when no other specific rule can be construed as governing a situation found during an OSHA inspection. Another reason is the interpretation that, if and when a serious injury or fatality does occur, it is virtually prima facie evidence that the employer did not provide employment and/or a place of employment that was free of hazards that could cause the accident. The result is the notion among operating management and safety professionals alike that they can never achieve perfection in their safety efforts, and therefore they are in a no-win situation against OSHA. We would like to counter that brand of defeatist logic.

First of all, rational people who allow that human fallibility is a fact of life do not, as a result, give up and do nothing to improve themselves or their real-life situation. When we finally decide in our heart of hearts that we will never be able to afford a Ferrari, we don't quit driving Tiempos. Conscientious managers and safety professionals should both be motivated by a desire to achieve the best that they can, within the resources they have to work with. Even while acknowledging an inherent inability to achieve a totally hazard-free work environment, both parties can work within their means to improve it.

Secondly, and most important from the parochial viewpoint of safety auditors, the general duty clause can be a powerful ally in selling management, even reluctant management, on the need for a specific hazard control measure. The task of auditors is simply to make rational and plausible accident or injury scenarios out of the documented conditions observed, showing that, as conditions are, accidents or injuries can happen. Therefore, prima facie, the employment and/or place of employment are not free of recognized hazards that can cause, or are causing, death or serious injury.

In Chapter 6, we used a vapor degreaser installation to illustrate the concept of using MCEs (maximum credible events) to target the appropriate hazards. To build on the slight familiarity we may have engendered regarding vapor degreasers, we will try to show how to use the general duty clause to improve the safety of an otherwise compliant degreaser installation. The applicable portion of the OSHA Rules and Regulations in 29 CFR 1910.94(d). It, in turn, refers the reader to ANSI standards on ventilation. If a safety auditor with normal time constraints exhausts his or her avenues of standards research and can find no specific requirement for emergency power for ventilation equipment (as we could not), but it is clear that a boil-over will occur if exhaust fan failure occurs, then the auditor can effectively tie the need for emergency power to a well-developed scenario for power failure.

What we usually find in such installations is that the heating source is steam and the condensing elements are coils with cold water circulation through them. As long as the cold water flows, there is no boil-over threat. But like as not, the cold water is circulated by an electric motor-driven pump, and the back-up exhaust fan is driven by an electric motor as well. Usually, both electric motors are also in the same power distribution block or motor control center. A failure in a system "upstream" of the control center will obviously shut down *both* motors at the same time. But the steam is not electrically powered, so it keeps on heating the solvent in the degreaser. With no emergency back-up power, a boil-over is inevitable unless something is done. The state is set for invoking the general duty clause.

The information that will likely be available to the safety auditor for developing the credibility of the hazard would include the past history of power failures, the percentage of time that the degreaser is operating, the numbers of personnel working in the proximate area of exposure, and maybe even engineering studies to validate the surmised exposure. Once the credibility of the hazard is established and known to operating management, the legal requirement to provide a safe workplace virtually requires that precautions against power failure during hot operation of the degreaser be taken. The safety auditor should be flexible enough to recognize that other solutions might also be equally acceptable, but he or she will probably have gotten the management and engineering attention really sought in the first place.

Chapter 8

Standards

Standards are, without a doubt, the safety auditors' staff of life. They are the "what should be's" in the operations he audits. They can be considered the rules of the operational game, against which the audits are conducted. The more extensive and specific the standards are, applicable to a given operation, the more credible will be any specific findings uncovered. This does not necessarily make the conduct of audits easier, or more mechanistic. The need for understanding operations, the nature of their attendant hazards, and the effectiveness of hazard controls is still central to thorough and credible audits. These fundamental understandings give meaning and legitimacy to the applicable standards.

We use the term "standard" here in its generic sense, to include legal, regulatory, company and advisory, consensus, or industry standards. Any authoritative resource that indicates "what should be" is considered a standard, regardless of whether it carries the force of law or not. We will treat each of these in turn in this chapter, summarizing their hierarchy, provisions, and how safety auditors should use them to promote their ends of enhancing safety in the workplace.

State and Federal OSHA

At the very top of the standards hierarchy are the Federal or State OSHA Rules and Regulations. Probably of equal stature are the Worker's Compensation laws, and the Equal Employment Opportunity laws under which most of us work, but those subjects are outside the intended scope of this book. Also, in selected industries, parallel rules and regulations have been promulgated that are that industry's OSHA equivalent, such as the railroad and mining industries.

As we stated in the previous chapter, OSHA Rules and Regulations have the force of law, and are not to be lightly ignored. For the sake of brevity and simplicity, we are going to hypothesize here that all State OSHA requirements are similar to each other and to the Federal OSHA requirements. At the same time, we acknowledge that there are differences, but we do not intend to explore those distinctions that might exist. Rather, we would point out that any approved State OSHA requirement must be at least as restrictive as its corresponding Federal provision. We recommend that the reader bear with us as we discuss this section from the Federal Rules and Regulations, then review our thoughts and ideas in light of the particular State OSHA provisions under which an organization might function.

Part A of 29 CFR 1910 is generally administrative in nature. It authenticates the legal status of the Code, provides some definitions, outlines protocols for issuing, changing or repeal of provisions in the Code, and gives an overview of the Code's applicability when other existing codes also exist. Perhaps of greatest direct interest to general industrial safety auditors is Part 1910.6, which reserves the right to incorporate by reference *any* other standards and invests in such standard the same force of law as the Code itself, when so incorporated. There is the caveat that advisory portions of such incorporated-by-reference standards remain advisory under the Code.

The significance of this subpart is that safety auditors can find that an otherwise nonbinding standard has become binding by virtue of incorporation by reference in some portion of the Code. They are then clearly given another mandatory, nonoptional standard that is authoritative when applicable.

Subpart B basically itemizes specific industries, with pre-existing federal standards, which standards are adopted into the OSHA Code, and extended to all employers, employees, and places of employment covered by the OSHAct of 1970. Again, construction standards (for example) not normally applicable to, say, a printing plant, may now have the force of law, by virtue of this subpart, in that printing plant, if construction activities are undertaken. Other specific industrial standards adopted and extended in Subpart B are ship building, ship repairing, shipbreaking, and longshoring and marine terminal operations. The incorporated standards, now part of 29 CFR 1910, are detailed from other Parts of Title 29—Labor.

Subpart C, 29 CFR 1910.20, covers all of the various aspects of an employee's right of access to this own exposure and medical records when there is the possibility of an industrial health hazard exposure. Definitions, record keeping requirements, and access protocols are detailed, among other topics. A very valuable reference is noted in appendix B to 1910.20, undoubtedly useful to those in the chemical and compounding industries. The Registry of Toxic Effects of Chemical Substances (RTECS) of the National Institute for Occupational Safety and Health (NIOSH) is issued annually by the Institute, and lists nearly 125,000 chemical materials, nearly 34,000 of which are individual chemicals, and over 90,000 of which are synonyms. As a preliminary indicator of a potential health hazard, this compilation is an excellent resource, particularly when Material Safety Data Sheets are not available, or as a "second opinion" when they are.

In Subpart D, 1910.21 through 1910.32, walking- and working-surface requirements are detailed. As is typical of the OSHA format, Paragraph .21 is a list of definitions, and we recommend reading the definitions that apply to the layout of an operation, to ensure that there is no confusion when applying the provisions to an audit finding, or in making the finding in the first place. General requirements follow in Paragraph .22, covering housekeeping; aisle clearances and markings; pit, tank, and ditch covering; and floor load capacity markings. Most of the remainder of Subpart D sets out detailed and specific requirements for the guarding of openings in floors and walls, requirements for fixed stairways, portable and fixed ladders, and scaffolds—both temporary and mobile. Paragraph .30 discusses other miscellaneous, though specific, installations. These paragraphs are extremely detailed, giving specifications and dimensions and strength requirements, design and installation criteria, personal protection requirements, etc. In typical industrial operations, safety auditors could spend days, perhaps weeks, poring over design drawings and specifications, taking measurements of widths and thicknesses and angles and heights to ferret out every possible element of compliance or noncompliance with these provisions of the Code. In many cases, these kinds of discrepancies do not relate to any Maximum Credible Event, so it may well be that an audit will evaluate walking and working surfaces visually, not quantitatively. Safety auditors and the cognizant management have to realize that priorities must be established where resources are, in fact, limited. Knowing generally what safe walking and working surfaces look like, and visually judging what is observed, may be sufficient treatment of this aspect, when viewed in the context of much more serious hazards that may be identified in the operation.

On the other hand, in an operation where tripping, falling, and slipping are indeed the most chronic or serious hazards (the MCE), a detailed and quantitative assessment might be warranted. This is a trade-off that safety professionals must make on the merits of the situation.

Subpart E, Paragraphs 1910.35 through 1910.40, deals with means of egress from fires and other emergencies in the workplace. Paragraph .35 provides definitions of pertinent terms relevant to the topic. Of particular interest to safety auditors are the definitions of low, high, and moderate hazard areas, because egress requirements vary depending upon which of these categories a particular operation or operating location fits. As in Subpart D on walking and working surfaces, a great deal of design and installation criteria are detailed in this section, in addition to capacity requirements, fire protection and maintenance, exit marking, and other dimensional details. Paragraph .38 outlines employee emergency and fire prevention plan requirements, to include egress training, personnel accountability, and emergency containment planning and training.

The entire Subpart E is quite redundant with the National Fire Protection Association (NFPA) Life Safety Code 101. It is our recommendation that both Subpart E and NFPA 101 be studied by safety auditors, so that, regardless of the nature of the operations audited, or the seriousness of the MCEs, adequate and compliant egress can be determined, or deficiencies readily identified. We say this because inadequate egress can be more than an individual hazard. It can be virtually an instantaneous mass hazard, especially in a labor-intensive operating area.

Subpart F (1910.66 through 1910.70) regulates powered platforms, manlifts, and vehicle-mounted work platforms. This section should be reviewed by safety auditors any time that the operation or area has occasion to use this type of equipment. Virtually any safety aspect concerning the design, operation, or maintenance of such equipment is covered in this Subpart.

Subpart G (1910.94 through 1910.100) deals with environmental controls, including ventilation, noise exposures, and radiation exposures. Paragraphs .94 on ventilation and .95 on noise hazards control are rather universal in scope. Several American National Standards Institute (ANSI) standards are referred to in the treatment of grinding, and in point-of-generation exhaust systems, including spray finishing installations. Lower Explosive Limit (lel) determination and control for flammable dusts and vapor-generating operations are also discussed. Again, virtually any design, operating, or maintenance hazard control criteria that one might imagine is covered in this paragraph.

Paragraph .95, on noise exposures, covers protection criteria, requires a hearing conservation program when exposures exceed those criteria, and outlines monitoring, employee notification, testing and protection requirements. Also, record keeping and retention requirements and employee access provisions are cited.

Paragraphs .96 and .97 on ionizing and non-ionizing radiation are based almost wholly on Title 10 CFR Part 20. Auditors conducting safety audits in these very specialized circumstances should refer to that source. We will not go further on this subject because of its highly specific and narrow applicability.

Subpart H (1910.101 through 1910.120) deals with hazardous materials. General requirements for compressed gases, cylinder inspections and relief device usage are covered in Paragraph .101, and requirements of the Compressed Gas Association are incorporated by reference. Paragraphs .102 through .105 deal specifically with acetylene, hydrogen, oxygen, and nitrous oxide. Only the paragraphs on hydrogen and oxygen provide independent requirements; the others refer to publications of the Compressed Gas Association. Certainly, safety auditors working in areas where these gases are produced or processed should become intimately acquainted with these OSHA sections and the subtler references that have been incorporated.

Paragraph .106 is a comprehensive regulation on flammable and combustible liquids, and will likely be one of the most used sections of OSHA law. Every conceivable aspect of these materials is treated, from tank and container specifications to spacing, diking, venting, electrical fire control and ventilation, static grounding, and transfer and maintenance activities. We suspect that there are a very few operating plants or areas that do not have some occasion to use or handle flammable and/or combustible liquids. We recommend learning the topical material included in this paragraph.

Paragraphs .107 through .111 deal with spray finishing operations, dip tanks, explosives, liquified petroleum gas storage and handling, and anhydrous ammonia storage and handling, all fairly narrow in their applications. Each is quite detailed in describing requirements for these specific operations. Safety auditors working in any of these operations should become familiar with the provisions pertaining to those operations.

Paragraph .120 is another that safety auditors will find applies almost universally, irrespective of what kind of industry they are auditing. This section outlines requirements for hazardous waste management and emergency response. We again recommend that safety auditors know this paragraph topically, and refer to it on virtually any safety audit where hazardous materials are being processed or manufactured.

We have made this OSHA review, to this point, in order to illustrate the content, scope, and applicability of the document. It has also shown that the OSHA Rules and Regulations is not a stand-alone document. OSHA has incorporated by reference numerous other standards and regulations. It is also clear by now that the various subparts have applications ranging from almost universal to that of a single segment of industry. We will only mention the remaining subparts topically and comment on their range of applicability.

Subparts I through L address personal protective equipment, environmental controls, medical and first aid availability, and fire protection respectively, and should have universal application. Subpart M is specific for compressed gas and air equipment, but will be applicable in most installations where a compressed air system is in use. Subpart N deals with material handling storage, covering vehicles, cranes and derricks, slings, and even helicopters. This section will find broad application in safety auditing. Subpart O on machinery and guarding will also be widely used, as will Subpart P on hand tools. Subpart Q is specific to welding, cutting, and brazing operations. Subpart R addresses several specific industries, including pulp and paper, textiles, bakeries, laundries, sawmilling, logging, agriculture, and telecommunications.

Subpart S on electrical will have universal application, while Subpart T relates to commercial diving operations. Finally, Subpart Z is the section on toxic and hazardous substances, and will be widely used by safety auditors, especially when the scope of their audits includes industrial hygiene matters. This subpart is extremely detailed, comprising about a third of the entire book of rules and regulations. It details system requirements such as medical recordkeeping, exposure data, and personal protective equipment fitting and testing, as well as highly detailed measures for numerous specific materials. Knowing the operations and materials involved will direct auditors to applicable portions of Subpart Z. However, unless industrial hygiene issues are specifically excluded from the safety auditor's charter, he or she should become familiar with the system requirements for any toxic or hazardous material exposure.

This overview of OSHA Rules and Regulations should assist safety auditors in ferreting out the safety requirements for an audit of an operation. The primary message we want to send is that they should be selective, but thorough, in determining which portions of the Code they use to audit against.

In-House Standards

These standards are the company's own rules of how its management plans to conduct its business. They can, of course, cover any and all areas of activity, from accounting practices to personnel and employee benefits, to purchasing, facility design, and (of course) safety. A viable operation cannot be managed efficiently in a chaotic environment, so some in-house standards, however terse or informal, written or not, must exist for any on-going operation.

Most companies that are reasonably modern in their way of doing business will have formalized their requirements into written statements of policy and implementing procedures. A policy on safety, and management's commitment to safety, should be at the head of a list of those in-house standards that safety auditors will extract as they plan their audit. Many of these will be quite general, though, and probably not give auditors anything specific to contrast with their findings on the floor.

Safety auditors should review all existing in-house management policies, procedures, directives, and standards, by whatever name they are known, to assess both the direction that they give to operating management and safety, and the adequacy and completeness of that direction. If safety responsibilities are properly focused on line management, auditors can assess whether the attendant authority and accountabilities accompany it. If all plant safety problems are decreed to be the Safety Department's problems, auditors will know how the situation got that way. This latter situation is, of course, unfortunate, but at least the auditors will know the culture they are entering.

More useful for the actual audit will be the operating procedures that are intended to specify how an operation is conducted. Often, these procedures will contain the safety precautions that the operating people are to employ for each step in the process, or each machine or material. Other, more formalized safety programs will have developed an in-house safety manual, or a set of rules and regulations tailored specifically to the company's or plant's perceived needs. This can even be extended to the individual operation level. For example, in a chemical plant, the safety requirements in a reactor area can be somewhat different from those of the warehouse or the shipping department. Safety auditors should become conversant with the content of the operating procedures, in whatever form they exist, that pertain to the operations they are planning to audit.

In so doing, they need to keep in mind two on-going assessments that they should be making. First, is the operation in compliance with the in-house procedures and standards? Second, are the in-house procedures and standards adequate and in compliance with OSHA and other regulatory standards? Findings that are negative to either of these questions must be reported. One of the most embarrassing findings from an OSHA inspection is to be told that you are not even following your own rules.

Advisory Standards and Codes

There are a number of reference materials that fall into this category. We include here unincorporated portions of the National Fire Protection Association (NFPA) standards, a fairly large number of American National Standards Institute (ANSI) publications, the American Conference of Governmental Industrial Hygienists

(ACGIH) consensus standards, plus innumerable local codes that may not, in fact, be advisory. These latter are beyond the scope of this work, but we would like to touch briefly on the others.

Depending on local codes and laws, many of the NFPA codes are advisory in nature. At the same time, they contain invaluable information and criteria for the safe installation and use of almost any imaginable item. A brief look through the index of the National Fire Codes shows standards for everything from AC grounding systems to boilers to ducting, care and use of fire hoses, incinerators, piping systems, and even zirconium milling operations. The master index, any section, or the entire code can be obtained from NFPA, Batterymarch Park, Quincy, MA 02269. We recommend that beginning safety auditors obtain at least the index, although the code itself would be preferable. With the index in hand, auditors will at least be able to order selected sections when questions arise.

Similarly, ANSI standard give a wealth of information on a wide range of topics. There are standards for agriculture, consumer products, electrical devices, fire protection, health hazard abatement and control, highway and traffic safety, nuclear installations, occupational safety and security matters, plus the ANSI/ASME boiler and pressure vessel code. A catalog listing all of the ANSI standards that relate to safety and health is available from ANSI, 1430 Broadway, NY 10018. We recommend obtaining this catalog and selecting those standards that fit the safety auditor's needs.

The ACGIH and the National Institute for Occupational Safety and Health (NIOSH) both publish recommended exposure limits for toxic and hazardous materials as consensus standards. These limits may not coincide with each other or with OSHA-adopted exposure limits for a given chemical. They are often more stringent than the OSHA limit, which obviously makes them more conservative. Often, OSHA and/or state OSHAs eventually adopt the more restrictive consensus standards.

The ACGIH, 6500 Glenway Ave., Building D-7, Cincinnati, OH 45211-4438, offers Catalog 1288, which lists numerous publications available from them. NIOSH also offers a Publications Catalog, from the U.S. Department of Health and Human Services, Public Health Service Center for Disease Control, in Cincinnati, OH 45226. The subject index can lead auditors, who are including health hazard evaluations in their audits, to publications on a great many specific material hazards, as well as methodology and identification and control requirements. Another publication we recommend from ACGIH, when specific exposures are to be evaluated, is their annual Threshold Limit Values — Biological Exposure Indices. The basic compilation and its annual updates are available at the same address given above.

Industry standards are time-honored resources to many. Such associations as API (American Petroleum Institute) and SAE (Society of Automotive Engineers) have adopted standards that touch on safety in those industries, as well as other aspects. Auditors who will function in those industries that have promulgated such standards would be well-advised to have the appropriate standards available.

Finally, we want to mention books and articles that abound in the safety literature. These are almost universally advisory, however authoritative or instructive they may be. However, they can be of considerable value, if not in identifying hazards, in offering useful solutions. Remember, the safety auditors' job should encompass both aspects. Constructive solutions from an advisory, but prestigious, source can help achieve that goal.

Contractual Standards

Especially in the aerospace industry, with which we are most familiar, and in many other governmental contracting industries, some safety requirements are usually defined in the contracts themselves. Undoubtedly, there are other segments of industry that function in a formally contracted environment — between government agencies and private enterprise, between separate governmental agencies, or in a regulated fashion between two or more private parties. We will not attempt to itemize examples of these, because of the almost infinite scope and variety that can be encountered.

Another common contractual relationship in operating industry is that between the employer and the employees — union contracts. Often, safety issues are spelled out in these contracts, and in fact that trend is increasing. We remember such occurrences as far back as the 1950s, specifically a union contract in which the company agreed to provide the needed personal protective equipment, and the bargaining unit agreed and promised to both use that equipment and enforce its use on its own members. While not a particularly hazardous plant, a rule was adopted that there would be universal wearing of hard hats. Had there been an OSHA in those days, hard hat requirements would not have applied in 95 percent of the plant. Yet a safety auditor, or an operating manager, or even the union steward would have considered it a contractual violation to go hatless in that plant.

In the governmental contracting and licensing industries, such as aerospace, nuclear, military and civilian procurement, transportation, construction and many other areas, safety auditors should make it a practice to find out from their employers or clients what those contracts cover in the way of safety requirements. In some cases, the contract will bind the operating organization to abide by other mandatory regulations, merely incorporating by reference such requirements such as NFPA, OSHA, DOT (Department of Transportation), or NRC (Nuclear Regulatory Commission) regulations, for example. In other contracts, specific regulations, manuals, and codes will be incorporated. The situation can easily arise where, in the absence of such an incorporated requirement in the contract, it would not apply to the operation in question. Safety auditors should find out what their contracts say, then ascertain what the referenced requirement provides, and determine how it fits into the operation they are auditing.

Deriving a Checklist

We have alluded briefly to checklists in previous chapters, emphasizing that they should never be viewed as a substitute for sound understanding, preparation, and planning, nor for keeping one's eyes and ears open while on the operating floor. We have now reached the point in the preparation for safety audits where a decision must be made. The question is whether safety auditors should use comprehensive (but generic) checklists, or should they now tailor them specifically for the operations they are planning to audit.

We have said previously that we do not recommend embarking on an audit without any checklist at all. We would liken that to beginning a trip to Bolivia (or Tibet, or Tanzania, or wherever) without a ticket, schedule, or map. Anyone with a memory as infallible as that should be in a profession much more profitable than safety auditing.

In Part 3, which follows, we have provided prototype checklists that we do recommend for general use. In perhaps 75-80 percent of the industrial operations that auditors will see, the checklists will serve as the memory joggers that they are meant to be. The questions that have to be asked are whether that is an acceptable level of confidence, and do the operations in question present such hazards that one cannot tolerate missing some important and unique aspect of the total exposure.

In order for safety auditors to adequately assess those questions, we recommend that a decision be made in conjunction with the reviews and "getting-acquainted" sequences developed in Chapters 5 through 7, and the first four sections of this chapter. While getting to know the hazards, considering the MCEs and prioritizing, and becoming familiar with the applicable requirements, they can make concurrent assessments as to whether there are unusual hazards and/or risks associated with a particular operation that would not be adequately addressed in executing the general checklist. At the same time, it will become apparent that there are probably numerous portions of the general checklist that will have no relevance to any particular operation about to be audited.

If auditors make notes as they identify these two kinds of differences, they can make their own decision of convenience as to whether to adapt the general checklist, or tailor one specifically for the audited operation. If there are no hazards/risks identified that the general checklist would not address to their satisfaction, they can tailor a trimmed-down version, or use the general checklist and leave some blank spaces. If there are only a few items that they determine need added evaluation, they can tailor a section addressing those, add it to the general checklist, and again, leave some blank spaces. Alternatively, if time allows, they may choose to create a tailored checklist for each audit, based on applicable portions of the general checklist and any areas that they feel need new or added emphasis.

As we said earlier, this is a personal choice of convenience to auditors. It is not nearly so important which choice is made, so long as nothing is overlooked. In Chapter 5, in introducing the idea of checklists, we recommended the use of a comprehensive checklist, emphasizing the folly of trying to conduct viable safety audits with no checklist at all. After leading the reader through the intricacies of the entire planning and preparation sequence, we will modify that recommendation slightly. Use the knowledge obtained in the planning and preparation steps to make intelligent decisions as to whether the general, comprehensive checklist is sufficient on all counts, or whether other areas of evaluation need to be developed.

If safety auditors are faced with developing, in whole or in part, some new items for a checklist, or creating one from scratch, the applicable requirements are the obvious starting point for the project. If it is clear that a requirement exists in law or policy, or that authoritatively good practice calls for a certain feature or practice, then it also defines a checkpoint for the checklist. Auditors can easily phrase a question from a provision in a requirement document. Also, it is much more feasible to tote a checklist around the plant than to carry all of the requirements documents themselves.

PART III

Effective Safety Audits

We have spent quite a bit of time thus far, and many pages, laying the foundation work for the stated purpose of this book. It is our conviction that effective safety audits must be conducted not only from the standpoint of compliance evaluation, but with a full understanding of the management environment, the hazards, and all of the requirements against which compliance will be measured. In addition, we believe that it takes far more than compliance with written standards to "make an operation safe." It is also management involvement and activity; the nurtured culture of the workers and managers; the understanding that they all have of the significance of what they individually do; and the style and reward standards that actually exist, not what is published in policy statement.

Machines, processes, and work methods may be immediate sources of hazards and accidents, even in the most enlightened management environment. How managers respond to identified hazards, before they cause accidents, and how they respond, or have responded, to past accidents after the fact, are vital elements of effective safety audits. True, complete audits look at the machines, processes, and work methods too. They may focus on physical hazards and exposures for the bulk of the expended time and effort. However, they will not be complete until they have interpreted those findings in terms of management performance.

PART III is our step-by-step roadmap for conducting effective and comprehensive safety audits. We will discuss the niceties of protocols and practices, management evaluation, physical plant/operations inspections, the assessment of work practices, and the vitally important aspect, through all of this, of communication. We will present our version of a comprehensive audit checklist, and make liberal use of examples in illustrating some of the concepts.

We will discuss special audits, such as hazard communication, fire protection, environmental or flammable solvent storage and control, in more detail, and talk about how to handle multidepartmental audits and surveys. While it is obviously not possible for us to depict, or instruct in, safety audits in every sector of industrial operations, we hope to offer enough, by way of generic material and examples, to get safety auditors on their way.

Chapter 9

Protocols and Practices

As in any other endeavor, there are good and not-so-good ways of undertaking safety audits. This chapter talks about some of the advised ways of conducting safety audits. The authoritarian approach is discarded, for good reason. The cooperative and factual approach is endorsed.

We will discuss the merits of the prior announcement of audits, as contrasted with surprise audits (we eschew the latter). We will walk novice safety auditors through the niceties of introducing themselves, their purpose and the nature of, and protocols of, true safety audits.

The importance of observing and reporting both positive and negative findings will be rehashed. Also, the methods and reasons for documenting those findings will be elucidated.

We will touch on the need for involving operating management in audits, comment on the extent of participation, and the active seeking of viable solutions. Finally, we will discuss the power of positive reinforcement—"catching them doing it right."

The conduct of safety audits can either be a confrontational exercise, or a cooperative attempt to identify and solve safety problems. Which alternative obtains is largely the responsibility of safety auditors and how they go about this acknowledged unpopular task.

Announced or Unannounced?

There are two schools of thought on the question of whether audits of any kind should be surprises or whether they should be known in advance by the auditees. One theory says that if an audit is well publicized in advance, the subject of the audit will go to great lengths to get his act together. The manager will do everything that was supposed to have been done all along. The flip side of the argument is that a surprise audit will give the auditor a truer picture of the "real world" conditions. After all, the auditor is supposed to contrast "what is" with "what should be." An unannounced audit will give the auditor a much more spontaneous introduction to the subject at hand—what is.

Which approach safety auditors take will depend to some degree on what their marching orders are, what *their* managements' expectations are. If their purpose is to catch the operating people and nail them to the wall, they will obviously not announce their audits in advance. But if the auditors, and their management sponsors, want to catch them doing it right, and cheer them on (as Peters and Waterman[1] put it), then they will give the operating management ample opportunity to prepare and "do it right."

In the first case, the auditors will find a situation that shows what operating management considers acceptable, or tolerable, or the best that they can do. In the second case, where operating *management* has had a chance to put their house in order, the auditors will find what operating management thinks looks good. After all, if the overall objective is the reduction of hazards and the enhancement of safety, and prior notice of an impending safety audit stimulates such activity, the goal is at least partially, or temporarily, achieved.

We have discussed, in previous chapters, our belief that the most effective safety activity is that of the operating managers. The role of any safety professional should be to guide, advise, teach, stimulate, and encourage that activity. It is, therefore, our opinion that safety audits should be announced ahead of time, so that any activity that the announcement generates takes place as soon as possible. This gives the auditor the opportunity to "catch them doing it right and cheer them on." We must weight the pros and cons. There are, in our opinion, more benefits to be gained from reinforcing what has been done, even the improvements we do not become aware of, than there are from "gotcha's" that get corrected later, after the audit. When positive measures are taken on behalf of improving

---

[1] *In Search of Excellence*, Thomas J. Peters and Robert H. Waterman, Jr. New York: Warner Books, 1982.

safety, and receive recognition in the audit report, the operating managers and their people will almost certainly be less defensive about other audit findings that need correcting.

Some administrative details that safety auditors should consider are the form of the notification, and the timing of it. It can be either verbal or written, and a verbal notification can be either by telephone, or in an informal personal visit. Of course, it can be both verbal and written. An informal verbal communication has the potential advantage, if done right, of giving the operating managers a sense of confidence that the auditors will give fair and objective assessments of the safety posture and the compliance status of the areas. It will allow the auditors to determine to some extent what type of managers they will be dealing with, if they do not already know. A nonadversarial relationship can be initiated, provided both individuals are willing to work toward that end.

We recommend that safety auditors make that initial contact in as unobtrusive a manner as they can. They should assure the operating managers of their objectivity and desire to maximize the safety of the managers' operation, but at the same time, make it clear that 'what is" will be reported. We also recommend that auditors follow that informal introduction with a formal letter or memorandum, that documents the intent, the operational scope of the audit, the approximate duration of the audit, and who, if other than themselves, the auditors will be. It is generally a good idea for safety auditors to document everything associated with their audits, so that they are prepared to answer any future questions on any aspect of the audits. More will be said a little later on documentation.

With regard to the amount of time that should elapse between initial notifications and the actual audits, we recommend one to three weeks. Too little prior notice approaches no notice at all. Too lengthy a notice period can result in its being forgotten, or preparation being put off until later. If the informal and written announcements are followed within a couple of weeks by the audit itself, operating managers will tend to treat them as current events, and therefore important — now.

We have included here a suggested prototype of an interoffice memorandum that might be used as a model for announcing an impending safety audit. As the reader can see, we do not consider it necessary, nor even advisable, to go into any detail. There is no need at this point to impose any prior constraints on the nature of the safety matters that might be investigated, or reported.

INTER-OFFICE MEMORANDUM

To: Operating Manager, ABC Department
From: Safety Auditor, Home Office
CC: President, Operations Vice President, Plant General Manager, Corporate Safety Director, etc.
Subject: Safety Audit of ABC Department
Date: August 20, 1991

    A comprehensive safety audit of ABC Department is planned for the month of September. I(we) will begin September 5, and expect our work in your department to be concluded by about September 22. I would like to hold an introductory meeting with you, and those of your staff that you feel are appropriate, on the morning of the 5th. I would appreciate it if you can arrange to have process drawings, equipment layouts, operating procedures, maintenance and training records, and other pertinent documentation available at that time.

    I would also ask that you designate one point of contact, if other than yourself, for a daily debriefing of findings, and to help answer questions as they arise. I am looking forward to working with you and your associates, and expect this project to result in significant benefits to us both.

(signed) Safety Auditor

Introductions and Intentions

Let's assume now that the safety auditor(s) have notified the operating manager of an upcoming audit a couple of weeks prior to the beginning of the audit. Let's further assume that the auditor(s) have spent those couple of weeks getting to know the operation, studying the paperwork, understanding the hazards, identifying the standards, both internal and external, and perhaps developing a checklist. The opening shots are about to be fired. Will they start a battle?

Hopefully not. The auditor(s) have a considerable amount of influence over whether the audit turns into a battle, or into a joint campaign for safety excellence. It is the auditors' show at this point.

It is within their power to initiate the interchange in a positive, cooperative vein. For example, suppose the requested documentation has not been assembled. Before scolding the operating manager about its absence, the auditor(s) should inquire as to whether the past few weeks or months have been quite busy. When they later ask about the requested documents, the auditor(s) will have already established an understanding that the manager might not have had time to gather them together. Of course, if the documents are not forthcoming rather quickly, then is time enough to exhibit a measure of annoyance. The point is—find a way to *not* start an interchange on a sour note.

Presumably, the operating manager will have gathered one or more of his key operating people, perhaps a foreman or journeyman, an administrative assistant or floor supervisor. The auditor(s) arrive at the appointed time, and everyone makes introductions. It is helpful if the auditor(s) reiterate their professional credentials in terms of employment with the company, or, if in a consulting role, to itemize some past auditing experience for somewhat similar clients. The point of giving that kind of a thumbnail sketch is to build some credibility at the outset. However, it cannot be done effectively if done with an air of haughtiness or in an authoritarian manner. That will only engender resentment. Also, the beginning of the introductory interviews may not necessarily be the most opportune time to cite those credentials. It may be more appropriate at a later point in the conversation, when a topic can be related to a previous experience. Back in Chapters 1 and 2, we talked about dealing with various kinds of management styles. The opening meeting on the manager's turf is the time and place to put those ideas into practice and find the right way to say what needs saying.

The auditor(s) should also restate their audit plan. The scope and duration of the audit, the standards that they have deduced to be applicable, and their own on-the-floor schedule should be covered. They should inquire once more what escort rules exist for visitors, what indoctrination or training is specified for nonresidents, and what personal protective equipment is required when they are on the floor. Also, their point of contact during their floor stay needs to be established, as well as a fairly rigid schedule for daily findings review. The auditor(s) should explain at this point that it is in the manager's best interest to have these daily "data dumps," because

(1) it will give the manager an opportunity to clarify or correct any misunderstandings that the auditor(s) have, and (2) it will give the manager a chance to take corrective action steps immediately for findings that are real.

It is also recommended that the auditor(s) restate their protocol of acknowledging timely and effective corrective actions taken or initiated during the course of the audit by so stating in their written report. To many operating managers, this reassurance will somewhat sweeten an otherwise slightly bitter pill. Of course, to others it may not be of any perceived consequence, but that issue is between the manager and upper management. It is worth the auditor's saying it (and of course, doing it), if it is helpful in allaying the operating manager's reservations, and it can do no harm if it doesn't.

If the auditor(s) plan to be present in the operating area on odd shifts or for a night shift or two, it is tactful to tell the manager up-front of those plans. If they intend to interview line workers about their work or procedures or perceptions, say so in this opening discussion. If they intend to delve in-depth into management systems and activities, and that is so stated in their formal letter announcing the audit, reiterate it now.

We suggest that it is only fair that the scope of the audit be consistent from the announcement to the introductory in-briefing, and through the audit itself. To change targets abruptly puts an undue burden on even a manager who is truly trying his or her best to help and to learn and improve his or her operation. If, on occasion, it is found that a serious situation does exist outside of the intended scope of the audit, the auditor(s) should treat that as a rare exception, and include the manager in the rationale that prompted the shift in direction or emphasis. Make the manager a party to the decision by educating him or her on the new problem. However, back at the initial in-briefing, those things are unforeseen, and the auditor(s) should simply state that the intended scope remains as described in the memorandum, and that any departure from that plan will be discussed with the manager before it is implemented.

The manager should be acquainted with the distribution that the formal audit report will receive, if he or she is not already. Depending on in-house customs, this will probably include the manager's chain of command up several levels, as well as the auditor's boss. In some organizations, the findings may be publicized quite broadly, where there are common safety concerns in many departments. The report distribution protocol should be decided on in advance as a policy matter by the auditing staff and the management that *they* serve.

Another policy decision that should be in place, is when and to whom the operating manager should submit corrective action plans and accomplishments. As we inferred earlier, this should be a issue between the manager and his upper management. However, organizational policy should spell this out unambiguously, both as to whom and by when. We suggest that the safety auditor(s) should be included on the distribution, because they are in a position to help management, and that upper management assess the adequacy of the actions taken or planned. The auditor(s) should, however, emphasize that the operating manager is not being put in the position of answering to, or having to satisfy them, the auditors.

Our recommendation is that the operating manager be directed by management policy to respond in writing within thirty days to the same people that are, under policy, on distribution for the original audit report. A policy statement like this takes the arbitrariness and variation out of the auditor/operating manager relationship from audit to audit, especially when the auditor(s) are fellow-employees. It also gives traceability and tracking. If the auditor(s) have received no corrective action report within thirty days, they can remind the manager of his or her policy responsibilities and assignment, without taking on the role of an enforcer.

Given that these policy provisions exist, or are created at the safety auditor's initiative, these provisions can be recited at the in-briefing with the same detachment. "These aren't my rules, Jack. They're the company's."

In summary, the objective of the opening session between auditor and manager is to establish ground rules that both can be comfortable with, and end it on an upbeat note. The time and place, and the participants, for the next get-together should be fixed, and the plan for that meeting understood. If it is to review procedures, if it is to inspect the plant area, or interview a cross section of workers, let the residents know what you plan to do next.

Findings — Positive and Negative

We would assume that, by now, the reader will have surmised that we are firm believers in fairness and objectivity, of course, but also in the value of positive treatment of audit findings. The question may be asked, and rightfully so—how can safety auditors find a condition or practice that is blatantly contrary to the most fundamental rule of safety, and then try to treat it positively? We will try to answer that question later in this section.

First, however, let's go back to the beginning. Obviously, auditors will likely find equipment hazards, potentially dangerous work practices, and conditions that increase the likelihood and/or the severity of injury and property damage. These will also vary in their degree of seriousness. Taken as a whole, these findings are the negatives. Without a doubt, they need corrective action.

To many, perhaps most, this body of findings *is* a safety audit. It has an almost exact parallel in financial auditing. If accounting irregularities are found in a financial audit, the news is (relatively) headlines to the organization, at least. Sometimes, it is news to the whole world. Bad news travels fast, and is generally the most newsworthy. So we become accustomed to seeing and hearing little else but bad news—negative findings. There is a good chance that the operating manager, upper management, and probably even the safety auditor's boss, expect that a safety audit report will be simply a recitation of numerous deficiencies and violations of codes and rules. We recommend that those expectations go unrealized. A safety audit is intended to contrast "what is" with "what should be," remember? Just because "what is" looks exactly like "what should be" does not make it any less of a finding. If this is the case, it should be reported just as factually as if it were totally out of compliance.

Positive findings are indeed more difficult to identify than blatant noncompliances. One way to ensure that at least some are found and reported is to consider what possible negative findings were not found. The very fact that they were absent is a positive finding. For example, if a machine shop is audited, and thirty machine tools all have proper guards in place, the auditor should not report no findings—there should be a report that thirty machines out of thirty in the shop were properly guarded. Or, if a safety auditor finds, in a chemical manufacturing operation, that every single worker and supervisor always wears gloves, goggles, and respirator, always cleans them at the end of the job, and always stores them correctly, an extremely positive finding has been identified.

If, on the other hand, two of the machines had their guards

removed, or if one of the operators had a habit of leaving his respirator dangling around his neck, the auditor has certainly identified one or two negative findings. At the same time, however, twenty-eight positive findings have been identified in the machine shop, and numerous positive findings of compliant workers in the chemical plant.

We think that it is important for beginning or veteran safety auditors to establish a reputation for "telling it like it is," in all of its aspects. They should inform the operating manager at the very beginning that this is what they will do and, equally important, they must then do it. When we get to the prototype checklists later in this part, we will illustrate how to follow through on that promise while on the floor doing the audit.

Have we answered the question posed earlier in this section — how do auditors treat noncompliant findings in a positive way? To reiterate, the noncompliance is indeed negative, but undoubtedly has a whole bevy of companion positive findings. All we are suggesting is that both be noted and reported. Even if *all* of the machines were unguarded in that machine shop, the auditor will probably find that the machine operators are quite faithful in wearing their gloves for chip removal and their safety glasses while operating the machines. We report both sets of findings, and likely identify the lack of guarding as a management system problem, and not as thirty unrelated deficiencies.

Documentation

One of the practices in which people in general and safety professionals, in particular, vary widely, is in the area of documentation. Many people think of documentation as the stuff with which we convict someone. This, of course, is not the safety auditor's purpose when doing a thorough job of documenting audits. Our reasons for suggesting thorough documentation are to crutch our otherwise fallible memories, not only from morning until afternoon, or from floor observation time until report writing time, but between this year's audit and next year's.

There are four levels, or sequences, we recommend for documenting audits — field notes, "stand-alone" notes, checklists, and the final report. The first of these is the field notes and interview notes that auditors make in real time—when the observation is made, or during the conversation. If you are like us, these will be pretty rough, incomprehensible to anyone other than yourself. Their value, though, is apparent. They consist of some key words or ideas that have a clear meaning when combined with the memory of the time and place in which they were made. After a full day on a shop floor making observations, auditors who took no notes may not remember that the third boring mill from the east door had an oil leak in back of it. But note-taking auditors will be able to recall it easily from an entry such as "oil on floor—3rd b. mill." Similarly, during conversational interviews, remembering a manager's or operator's opinion or perception on a subject, when ten or twenty topics may have been discussed, can tax the most remarkable of memories. However, when that individual tells the auditor that cutting costs is his or her top priority, and that is where the majority of that person's time is spent, a note like "cut cost-#1" will be crystal clear later on when the auditor translates his or her field and interview notes into the second level of documentation—"stand-alone notes."

We use this term simply to indicate that this level of finding documentation is unambiguous by itself. A stand-alone note, or page of notes, is dated, identifies the operation/location being audited and the auditor, and describes each finding (both positive and negative) in such a manner that another person familiar with the area could go to the finding and see it too. Stand-alone notes may be hand-written, like field notes; it's not important because they will not leave the audit file. Their purpose is to capture the finding in the detail necessary to understand and envision it days and weeks later when the report is being written. So the field note that the auditor scratched out about the oil spill behind the boring mill might be rewritten as:

12/10/88 Machine shop, Building 21   Auditor Jack Jones

1. An oil leak from boring mill no. 3 has caused a puddle in back of the machine. Slip hazard.
2. Etc. Maybe finding 2. is that 29 machines had no oil leaks or puddles.
3. Etc., etc., etc.

Stand-alone notes should be accumulated throughout the audit cycle until all observations and interviews are considered to be complete, or at least no more are planned. This is the time to start filling out the checklist(s). Each stand-alone note is either a positive or negative finding. Since a checklist entry, such as housekeeping, has numerous opportunities to be an observation, the checklist should be used as a kind of a scorecard, where all the positive observations concerning housekeeping are counted, and all of the negative observations are counted separately. This enumeration process then gives the auditor a basis for reaching an overall assessment of the status and attention given to housekeeping in that operating department. One oil leak, unmopped and unrepaired, is one negative finding, and may reflect on one machine operator's view of its importance. But twenty-nine clean machines is probably a better measure of the operating manager's emphasis on housekeeping. It is the total picture that the auditor wants to depict accurately when reporting on housekeeping, not just the leaker.

Finally, the published report, the final product of the audit, is the fourth level of documentation. Writing skills are a major asset to safety auditors, because their end product almost always has a high-level readership in the organization, and it has an aura of permanence about it. Besides, their reputation and acceptance are involved, and the more professional the report is, the greater is their perceived credibility. We will discuss the final report in more detail later in this chapter.

One other protocol that we have mentioned before in the general area of documentation is citing the source document that defines an observation as a finding. When the oil leak is put in the report as a negative finding, cite 29 CFR 1910.22(2), and all argument about whether that is a hazard will probably disappear. The hazard has now been documented.

As we said earlier, a year later, or five years later, when it could become important to know whether some hazard had occurred before, and whether correction was made, the auditor can produce at trail all the way back to the original observation, and there will never be a question as to its authenticity. Audits have a way of resurfacing at the strangest times.

Eliciting Solutions

We have previously discussed the need to establish the credibility of both auditors and their findings. The surest way to establish the credibility of a finding is to cite the requirement, chapter, and verse. A rational operating manager can hardly reject out of hand a finding that is a clear violation of either a legal requirement or a dictum of the company's upper management. It is almost as hard to ignore expert advice from a nonregulatory source, such as ACGIH or NFPA.

The next step in effecting safety improvement is as much of a challenge as the first for auditors. This is to get managers to provide, or contribute to, the solutions—to buy into the corrective action. It should be *their* corrective action that is implemented, not yours.

To proactive operating managers who have integrated safety into their professional way of life, this will be automatic. They will accept the negative findings as conditions that must be changed to the positive, and lead out in solving, planning, and acting. In this best of all possible worlds, the safety professionals' role should be one of encouragement, applause, support, and advice.

Unfortunately, not all managers are like that. In the early chapters of this book we discussed the various reactions that safety auditors can anticipate to negative audit findings. They might be argumentative, oblivious, annoyed, defensive, or disinterested. The Auditors' job is to try to get past those stonewalling tactics, and stimulate some constructive thinking on the reluctant managers' part.

One helpful mechanism is immediately available in the auditors' own notebooks, if they have been making positive observations, along with the negative ones. They can put the negative findings into a positive perspective by pointing out how frequently the operation was in compliance, how these negative findings are so inconsistent with the general excellence otherwise observed. They might try to convince the managers that they may as well be 100 percent compliant as 97 percent.

If that approach is inappropriate (because the operation really wasn't very compliant), the auditors may have to be more prosaic about conveying the findings, simply stating that "these are your noncompliances, and we need to work together toward finding the most efficient way of getting into compliance." Remember, we defined safety auditors as pragmatic realists. A circumstance like this one is the time for pragmatism.

In any case, the offer to "work together" should be made. If it seems propitious, the auditors might also suggest that ideas be solicited from line workers and foremen, people who are closest to the problem and likely most aware of what *can* be done. It is almost axiomatic that people who do a task know the most about how to do it—both the right and wrong ways. People who run a machine know more than anyone how that machine works, or doesn't work. We believe that some of the most valuable and viable solutions come from this source.

There is a word of caution to be mentioned here, however. When auditors, or managers, ask for help from the line workers, they are often asking them to disclose something negative about the operation. Otherwise, a solution would not be needed. Disclosure that is punished, however subtly or subconsciously, will dry up that resource immediately. Conversely, disclosure that is used in a positive and productive way will encourage more disclosure. It will open communication channels so that very subtle hazards and problems will be brought forward, probably accompanied by solutions. Auditors should try to determine beforehand that managers will not use information from the operating people in a punitive or derogatory manner. If they cannot have that confidence, it is probably best to back off to working one-on-one with managers in reaching a mutually suitable solution.

The main goal in this aspect of auditing is for safety auditors to function as facilitators and advisors, not as accusers, not as a tattlers, and not as a director. That type of cooperative problem solving will give the manager, and his crews, a sense of ownership in whatever corrective action plan comes out of the solution seeking interchange.

Take the oil spill behind boring mill no. 3, used as an example in the last section. By citing 29 CFR 1910.22(2), we should have moved the problem beyond opinion and nuisance into the realm of an actual violation. Hopefully, the shop manager now realizes that doing nothing is not an option. The auditor may feel that the best solution is to take the mill out of service, tear it down, find the problem, and repair it—NOW. However, we propose that the auditor not take that position. The auditors initial report of the finding should consist of (1) reiterating the observation, (2) citing the OSHA reference, and (3) offering the suggestion that one obvious solution is to shut down and repair.

This makes it possible for the manager to acknowledge that something must be done about that persistent leak, while leaving the door open to other courses of action. It well might be that there are compelling reasons why the mill should continue operations for another day or week. The auditor should find out what they are, when the manager is reluctant to shut down now.

Here is where the auditor can come through as a facilitator, because it is clear that the oil leak is a slipping hazard, not that the machine is going to disintegrate and injure someone if it is not repaired immediately. The auditor can work with the manager to find other less disruptive, solutions. Again, rather than tell the manager, the auditor should ask how the hazard can be contained until a more opportune time for repairs. The manager might come up with rerouting traffic and barricading the area of the leak. The manager might impose an hourly inspection and changing of the absorbant used to soak up the oil. If these sound like reasonable intermediate fixes, the safety auditor should (1) accept them, (2) push for a time when shutdown and repairs will be done, and (3) follow up in a day or so to see if the agreement has been implemented.

The chances are high that the auditor will find the area barricaded and that the mill will be taken out of service when the manager said it would. The overriding reason will be that this was the manager's plan of action — not the safety auditor's.

Reinforcement

Many examples can be found in the current literature on the practice and value of "positive reinforcement." To the skeptic, the idea is trite. To the believer, it is his most valued tool for conducting interpersonal relationships. For those who have not heard of it, or not paid any attention, it has little meaning in terms of their daily activities. Positive reinforcement is a cornerstone of Dan Petersen's approach to safety and safety management[2]. Our definition is not

---

[2]*Safety Management*, Dan Petersen. Goshen, NY: Aloray Inc., 1988.

original and is not especially elegant. For safety auditors, we suggest that positive reinforcement is the business of "catching them doing it right— and rewarding it." The reward idea may sound a little presumptuous, but we are talking about reward in its broader sense —an acknowledgment, a word of appreciation, or a specific audit report finding, not a prize.

The usefulness of positive reinforcement can be apparent on at least two levels during an audit. First of all, it will promote a friendly, interactive relationship with the line workers that the auditors are observing. Even if the audits begin in an atmosphere of distrust and suspicion, a few strokes are bound to soften up that uncomfortable start. If the workers perceive that the rewarded act or condition is the correct one, it will eventually effect their resolve to do it that way. A good lead-in to commenting on a positive observation is to say something like, "that was exactly by the book, Charlie. Do you know why the safety rules say to do it that way?" If Charlie does in fact know the why, another compliment is in order. If he doesn't, the opportunity opens up to teach him why.

By the same token, positive reinforcement can be a very effective and persuasive tool in dealing with the operating manager, who is usually human, too. All of us, whether we admit it or not, tend to bask just a little bit in the presence of an accolade that is clearly directed at us. Even the most taciturn manager is not totally immune to that slight glow that accompanies a compliment. When twenty-nine out of thirty machines are found to be in good repair, and the auditor says, "You must be doing it right, Jim—twenty nine of your machines were in excellent repair," it implies that Jim is taking an active part in making sure that his machines *are* in good repair, and will encourage him to do it again. So what if Jim never bothered about machine maintenance before, if he starts bothering about it now? The objective is improved safety and safety performance, not to nail Jim for a machine that was in a deteriorated condition. The negative finding will tell its own tale in different way. Meantime, a step has been taken in the right direction.

Other opportunities will probably arise for positive reinforcement during the course of safety audits. Beginning auditors will soon learn that many negative findings can be corrected almost immediately. Most notably, housekeeping messes can be cleaned up in minutes, or hours, at most. Flammables can be returned to the proper storage cabinets, an open high-pressure air hose can be fitted with a regulating nozzle, a guard can be reinstalled on a machine, etc. When immediate corrective action is effected, or when an immediate start is made on an action that will obviously take days or weeks to complete, auditors can reinforce such promptness with their own appreciative response, and perhaps more important (to the responsive manager), the auditors can — in fact, must — acknowledge that prompt action in their audit reports. The bottom line in this endorsement of positive reinforcement is simply this. It will never discourage or defeat safety improvement, and it probably will promote it. Even if it doesn't, it certainly can't hurt, and it will make for a more pleasant working environment.

Chapter 10

Evaluating Management

To some, this may sound like a terribly presumptuous thing for safety professionals to undertake. After all, operating management directs the activities that build or make the products that provide the enterprise with the income that pays their salaries. In addition, the traditional role of safety professionals is to see that the storage cabinets are neat and clean, that the wiring and plumbing are to code, and that safety shoes and glasses are being used. This has, in the past, freed managers to do their thing—manage their operations.

However, more enlightened thinking has been brought to bear in recent years, most succinctly phrased by Dan Petersen. The job of operating management is "safe production," not just production. It is from this thesis that we say that safety professionals have a right and moral obligation to evaluate management's proficiency in safety, just as accounting professionals measure their economic performance, and quality professionals determine their reject rates. Even the Sales Department gets in on evaluating operating management's performance, in terms of on-time deliveries, and the Human Resources, or Personnel Department will likely measure their performance in terms of complaints and grievances.

So it should not be, though to some it will be, a revolutionary idea for safety auditors to audit management's safety performance. We have compartmentalized this evaluation process into five headings —involvement and commitment, activities, perceptions, systems, and delegation. However, these titular distinctions are rather imprecise, and really roll up into an assessment of the manager's ability to manage and direct "safe production." We will try to provide measurement yardsticks by which safety auditors can make valid determinations—the "standards" for management's safety performance. Clearly, these standards are "advisory," in the context of Chapter 8. We suggest that they be reviewed with upper management and adopted in whole or in part by the safety auditors' sponsors, before being applied in audits.

As we have indicated on several occasions before, and will touch on again, there is nothing quite so fatal to the auditors' credibility as citing a deficiency for which there is no requirement. If there are no management safety performance standards in the organization being audited, our firm advice is—have some standards established and audit to them. Do not audit in a vacuum.

We feel that this chapter does give the reader a fair picture of what proactive management looks like, however, even if it is not an organizational standard or goal. The probability is high that, lacking safety performance standards, managers who function something like those that we describe here, will also stand out in terms of high productivity and crew morale, as well as low accident rates. In the end, that correlation may help institutionalize the proactive practices of these managers.

Measuring Involvement and Commitment

One person says of another, "this person is really involved in the job," and another says, "that person just doesn't want to get involved in this problem." What is the basis for these observations? How did these observers reach such seemingly unambiguous conclusions? The chances are such statements are the subjective summation of what the person in question has said and done, the time and energy that the person has been seen expending on an issue, the person's facial expressions when the issue comes up, even the person's body language when talking about it.

Auditing, however, and safety auditing in particular, is a viable discipline only when approached objectively. Subjective evaluations of housekeeping, hazard communication, or pressure relief systems will not be very credible, and neither will subjective assessments of management's performance in their safety responsibilities. Imagine the response of an operating manager or a plant general manager when an auditor reports, "I would like to see two lock-out switches for that machine, because that will make it safer than one during machine maintenance." The manager is very likely going to ignore the auditor. The auditor will also be ignored, probably even ridiculed, if he or she reports that an operating manager is "not sufficiently involved," and lacks any data to confirm it.

So we need to understand how one observes and measures safety involvement. We have found that the most accurate way is to accompany the manager throughout the day, and for a number of days. Interviews and questionnaires simply do not give a true picture of the manager's routine. It takes observation, and the recording of each instance of a safety concern or question, or expression of awareness, that the manager exhibits. This is not so much a measure of the manager's competence as a safety engineer, as it is a measure of the manager's recognition that safety is a prerequisite to success.

It may seem awkward at first, following the manager around all day, but safety auditing itself might be a bit awkward at first, too. After a while, both the manager and the safety auditor will become acclimated to the situation. The objective for the auditor is to see the manager's way of managing the operation in real time, and in the real world of day-to-day production. If the manager has a daily or weekly production planning meeting with the operation's foremen and quality inspectors, or with the purchasing, shipping, or material control people, the safety auditor should attend. If the manager has a one-on-one stand-up problem-solving conversation with one of the operation's foremen, the safety auditor should listen in. The manager tours the operating area, talks to line workers, interviews prospective employees, works over the operation's budget and production figures, or meets with the boss's staff, the safety auditor should be there. The manager's involvement cannot be ascertained from memos and statistics.

At the same time, we recognize that there will be varying degrees of acceptance of this shadowing technique on the part of operating managers. There will also be those few items of management chores that are inappropriate for observation by auditors, such as counseling an employee, distributing merit raises, etc. By and large, however, a manager should not object to doing operations planning and directing in the open. If the manager does object, we suggest that that fact is a finding too. We will discuss that point a little later.

It is not inappropriate in this phase of the audit for the auditor to ask questions. After all, the subject of an audit must expect to be talked to and questioned. When the manager makes some decision, the auditor should feel free to ask what factors entered into the decision-making process. When the manager walks through the plant, the auditor can ask what is being looked for, and what the manager sees of significance. In meetings, it is probably best to save the questions for afterwards, unless the auditor finds him or herself a welcomed participant in the meeting. If the manager spends an hour or two wrestling with a budget or production schedule, or planning the installing of some new equipment, the auditor should follow up afterward to ask what was decided, what factors were weighed, and which ones drove the decision in the direction that it went.

We suggest that the operating manager's safety involvement is measured by two counts—how often safety is a consideration in the manager's everyday world, and how often it is the determinant. The counts are made as the auditor shadows the manager over a period of days or weeks. The important point is to understand when safety *should* be a consideration, and should be decisive, as well as how often it was. We suggest also that there are extremely few occasions when an operating decision, plan, or action has no relationship to safety considerations.

Every operation and operational entity is different, so we are not going to try to be specific about what is "par" for these counts of safety involvement. Some managers are planning and making decisions every few minutes, while others may face them only a couple of times a day. The more accurate measure is the percentage of times that safety was in the manager's mind when it should have been, and the percentage of time that it was a major factor in the equation.

Another clearly negative measure is the number of occasions when plans and actions are implemented by ignoring, or overriding a safety concern that has been considered and understood. When the manager, for example, asks whether a hoist can handle a certain load (because the safety implications of a hoist failure have been considered), and is told by the structural engineer that it will not safely handle that load, but then instructs the work crew to lift it anyway, the manager has demonstrated how committed he or she is to safe production.

We look at involvement and commitment as two sides of the same coin. People are often involved in something without being committed. Businessmen join clubs, safety engineers join ASSE, not necessarily out of commitment, but because it is the thing to do. Managers become involved in and with safety because it is the current wave, or upper management has decreed that they will do so. They can learn a lot about it, and bring it up at all the right times. The true measure of commitment is how faithfully they adhere to the known requirements when push comes to shove.

We alluded earlier to the probability that some managers will not be overjoyed at the prospect of managing their operations in the bright glare of sunlight. In other words, they may not permit the auditor to observe their management methods and processes, at least in their entirety. This in and of itself says a lot about such managers. Our advice, however, is to not make judgments about motives, nor to argue with the managers about why they should or must cooperate. Having explained that upper management has approved the protocol for evaluating involvement and commitment, and how it is to be done, auditors, should back away from any further confrontation at this point. They should make what observations they can, and then report on them. One of the observations will be a factual statement that the operating manager declined to allow the auditor to accompany the manager in some of the activities and deliberations.

As stated in the introduction to this chapter, safety auditors should not embark on a management evaluation protocol until the topical material, the "standards," have been worked over with, and approved by, upper management. If the latter have no interest in the matter, safety auditors will have a very difficult time, and little influence on, improving safety in the plant through operating management.

Management Activities

In Chapter 1, while discussing proactive, reactive, and passive management styles, we touched on some of the activities that truly proactive operating managers would do in integrating safety into their operations (see Table 1.1). Hopefully, we have explained and shown by now that a proactive management style is far more productive and efficient than the alternatives. Also, it should be clear at this point that safety is at its maximum when it is built into operations, from the planning stage right on through the design and layout of the operating facility, process planning and development, and into production. We have called this culture integrated safety.

Since we want to develop proactive managers, we need to be able to not only get a handle on their involvement and commitment—that is, their awareness—but also on the activities—the observable things that they can do to exercise that proactive style. If we can define proactive management in terms of things that can be observed, then we can measure, at least relatively, how proactive they are. So our purpose here is to expand on the items in Table 1.1, and discuss ways of measuring them.

We do not suggest that this will provide an absolute score, such that a bakery manager in Seattle can be compared to a printing plant manager in Cincinnati. Different industries will need and permit differing levels of activity, with differing emphases. But we do suggest that, in a given plant, this approach will distinguish the degree of proactivity among a number of second- and third-level managers. Furthermore, it will track progress and improvement, or regression, of an individual manager.

In Table 1.1 in Chapter 1, the first characteristic that we itemized under the heading "Understand and Prioritize Hazards" was the idea of studying and learning about the hazards of the operations for which the managers under evaluation are responsible. We used the verbs reviews, questions, studies, and learns, and the point was to provide descriptions of actions that one could either see directly, or see the results of. Secondly, we suggested that out of such basic understanding came the ranking of those hazards in order of priority. This means that the operating managers have determined in their own minds which hazards *must* be adequately controlled, and which carry lower levels of urgency. Our thesis is that if managers have hazard rankings at their finger-tips or hazard control plans for their operations, and can recite the control status for each item, then they have demonstrated a proactive initiative in this aspect. If, on the other hand, they have to call in their foremen, or the shop planners, or process engineering for help on such a topical subject, or if their answers to questions are vague or superficial, we would not give them positive evaluations on their understanding and control of their operational hazards.

The other two items in this first section of Table 1.1 have to do with their more technical role in determining and accepting what resources they have to manage their operations. These are the facilities, tools, and processes or procedures that are at the managers' disposal. Proactive managers will have documented their approval of such basic resources. They may have signed off on the drawings. They should have approved the process flow or plan, the layout, and the detailed procedures that will eventually govern how the operations are run. They also should be able to show, if appropriate, how and when they documented safety problems, and required changes in procedures or designs in order to alleviate the problems. Managers who can document their evaluation and ultimate approval of the nuts and bolts of their operational resources and methods are proactive managers. If they have delegated these responsibilities to their foremen, or abdicated them to the process, manufacturing, and design engineers, they have not demonstrated proactive management in this area. This determination by safety auditors should not be construed as a measure of the actual safety of the operation, nor of the competence of the managers' reviews and approvals. It is a measure of their activity.

Part 2 of Table 1.1 deals with the managers' planning for safe production. We mentioned the review and assessment of the adequacy of operational controls, and the identification and correction of deficiencies arising from those reviews. The assumption was that operating managers have, and should chair, regular operational planning meetings with their staffs and appropriate support people. Ideally, all relevant issues are brought to these meetings, so that the managers have all of the information needed or available to make optimum production plans. Obviously, they have to know what products are needed, in what quantities and by when. They have to know what materials, manpower, and production equipment and capacity are available, and when. They should want to know the maintenance status of their equipment, what open, incomplete or unresolved safety problems exist, and who is or is not trained and qualified for the work. In other words, proactive managers will demonstrate, in this, the most basic facet of their job, the total integration of safety into their operational planning. Finally, knowing all of these facts and statuses, they will make conscious decisions as to whether any or all of the operations should proceed, what prerequisite actions are necessary before proceeding, or what modified plans will accommodate any of the constraining problems—including known safety problems. This proactive approach to planning safe production can be readily observed by safety auditors who can sit in on such planing sessions. We certainly urge that this protocol be written into the management authorization of safety auditing programs, and that it be exercised.

Part 3 of Table 1.1 concerned the managers' on-the-floor activities. These are the most observable elements of an activities assessment. The managers either do or do not have MBWA (management by walking around) schedules, and they either have or do not have documentation of such activity. The documentation may be in the form of notes, memos, or work orders. If they can produce their schedules and reasonable documentation, they have again demonstrated proactive roles through their MBWA activities. We recommend that safety auditors accompany the managers on a few of their MBWA tours, perhaps making their own notes to hand to the managers in a spirit of assistance. If the managers accept such input appreciatively, the auditors can surmise that they are sincere in their efforts toward proactivity. Given that the record of MBWA is there, auditors can readily ascertan whether they are doing a thorough job of it, inspecting and communicating in all work areas of their operations.

We feel rather strongly that an essential aspect of the managers' on-the-floor regimen is thoroughness in the long term. From the standpoint of safety itself, no facet or work station should be omitted. From the workers' viewpoint, favoritism will defeat the overall benefits in morale that contact with them will inevitably generate. As we suggest in the Activities Checklist (Table 1.1), the managers should cover their entire area, they should cover each area in succession, in depth, and they should go on tours in both busy and slack times. If this routine is faithfully followed, they will not be in danger of (1) missing something important, or (2) building resentment because they only chat with their cronies.

It is important for managers to be thorough in their coverage. It is equally important for them to be thorough in their observations. Auditors should ask to see their notes or records from these MBWA visits. The auditors should look for both positive and negative notations. If there are none of the former, a warning flag is immediately raised, because the auditors are not dealing with proactive managers at heart. They are dealing with reactive ones. The next question should be whether the managers found nothing positive, or whether they simply verbalized their positive reinforcement.

A very telling clue to the operating managers' proactivity in their MBWA regimen is the depth of their observations and whether they seek out "root causes" of problems. If they can show an auditable trail from their floor notes through investigative follow-up to root causes and the correction of them, then the auditors have made a very positive measure of the managers' proactivity. Of course, if the observations are trite, with little or no substance to the corrective action, a fairly dismal measure must be concluded.

Communication with upper management is another measure of managers' activity in their safety responsibilities. For reasons that have, over the years, eluded us, some managers are extremely reticent about telling their superiors about safety problems. There may be a fear that it will reflect on their own competence. It may stem from a cultural bias in the organization that relegates safety issues to the Safety Department, so that management can forget them. Alternatively, it might simply be that the managers are oblivious and cannot discriminate between the important and trivial in safety matters. For whatever reason, the lack of upward communication, when higher level resources are needed to correct hazards, has proven to be an often tragic mistake. Auditors should ask for documentation that problems beyond the operating managers' resources and authority to correct have been evaluated. Whether actually corrected or not, that documentation will identify such managers as those who actively pursue their role in safety.

Part 4 of Table 1.1 has to do with coaching. By coaching, we mean simply how instructive the managers are when they find conditions or procedures that are different from what should be, or from what they want and expect. The first item in that Part is critical—that the managers explain to their people what is expected. No amount of correction and improvement alone can communicate to operating people what the goal is. The correction and improvement have to fit into a previously communicated and understood pattern of expectations. Any individual safety item, for which the managers direct change, will result (at best) in an isolated and probably begrudging implementation, if the managers have not created and publicized their requirements. Those directed changes that *are* logically in line with the managers' requirements will become much more widely adopted and adhered to.

Given that the managers do communicate their expectations, it is of equal importance that they explain why particular situations or practices (negative findings) are not acceptable, and how to correct them. This is usually a case of reiterating the expectations previously explained, in terms of the individual conditions that have been found, not an overly complicated concept. Relating the findings to their requirements will reinforce the requirements' legitimacy in the eyes of the managers' staffs.

This phase of management evaluation by observation can be accomplished best by far if the auditors can accompany the managers on some of their rounds. The next best measure of the managers' coaching prowess is to use the perception survey discussed in the next section. Suffice it to say that the perception survey will inform the auditors of how the coachees regard the coachers—that is, how the operating personnel grade the way in which the managers inform and instruct them on safety issues and deficiencies. A poor third, in terms of measuring coaching proficiency, is interviewing the managers themselves. They can hardly be objective, and may likely tend to either understate, or overestimate, their effectiveness.

However, in an on-the-floor setting, the auditors get the best basis for this evaluation. If all three elements are present (clearly articulated expectations, explanations for how situations are deficient, and what corrections are needed), and they are offered in a teaching, nonaccusatory manner, the managers' proactive stance is proven, as it relates to coaching. Again, it has not addressed how thorough or competent the managers' coaching has been, nor how accurately they have identified deficiencies needing coaching and correction. But it has evaluated their proactive coaching skills.

Motivation is the subject of the fifth part of Table 1.1 It is formulated primarily from the positive perspective. It is a truism that we can only motivate ourselves; we cannot motivate others. We think that this is semantic gamesmanship, in that we *can* provide widely varying inducements and rewards with which others may motivate or demotivate themselves. Be that as it may, numerous experts in behavioral science (too many to enumerate) have studied the mechanics and results of positive feedback and behavioral modification. The idea, previously touched on in this work, is to "catch them doing it right, and reward it." In other words, a manager should make it a pleasant experience for a worker to be observed following the instructions that have been given.

Managers should do this, not with slogans or contests, or with anonymous nominations from their foremen, but one-on-one with each and every individual on their crews. If the managers are truly involved in MBWA, their notes will reflect their observations of "doing it right," and who was doing it right. Auditors should ask for a list of safe performance observations by individual, who received special recognition for it, and what that recognition was. Again, a record of the managers' activities can be a valuable source of information for auditors trying to evaluate the manager' proactivity. Perception surveys can also be a valid measure of the managers' positive motivation methods.

The question of how to judge the managers' handling of negative observations naturally arises. They should have some, and they should want to demotivate those practices. They should—and must—correct them, of course. Managers certainly cannot provide positive feedback and motivation for such practices. Their proper course is to revert to their coaching skills and reiterate what they want, why and how, and provide no other feedback. The effected operators (or foremen) then have no reasons to feel "picked on" or chastised. They have only been reinstructed. It will soon become apparent, if the managers are fair, thorough, and consistent in their observations, instructions, and recognition, that doing it right is really what the managers want and reward.

Part 6 of Table 1.1 may be a radical new idea to some—even some safety professionals—who have grownup in the era where, once an accident has happened, it is up to the Safety Department to figure out why and what to do about it. However, if safety is to be an integrated management function, we suggest that investigation of the failures—the accidents—is a basic part of that function. Proactive managers will want to find out what went wrong, and will enlist whatever help they need, including the Safety Department, to find out. They will use Safety as a resource, not as a crutch. It will be their initiative to determine root causes, not only immediate, or proximate ones, because they want to know that the corrective action will treat the problem, not the result. Also, the managers will document what they find, as well as what corrective actions they take. Auditors should look at the documentation, then at the operations, to evaluate the appropriateness of the corrective actions, and how vigorously they were implemented.

Finally, we inquire into the operating managers' follow-through practices, as described in Part 7 of Table 1.1 As we all probably know only too well, the most intelligent and elegant solutions are often lost to a fallible memory. If there is no method of jogging that memory back to what has slipped away, our best intentions are for nothing (that is why safety auditors should use checklists). Operating managers should have some method in place that is their memory jogger. It is not only for safety hazard corrective measures. Working action item tracking systems are very useful for any and all types of detailed follow-ups.

Safety auditors are not really concerned about how widely applied the managers' tracking systems are. The main interest is whether they are used for ensuring follow-up on safety matters. Review of the systems, and some of the data in them, will answer the question of whether the managers are proactive in tracking corrective actions through to completion. The second aspect of corrective action follow-up is in what the managers do after their status tracking indicates an unacceptable status or progress. This is a bit more subjective than most of the other items we have discussed in this section, but doing nothing is objectively distinguishable from doing something, and that is what the safety auditors should look for. This measure of proactivity (doing something to get a corrective action into place) does not evaluate the correctness of the something. It does measure proactivity.

We have introduced a fairly extensive number of ideas in this section on measuring managers' activities, to evaluate their style of management. We may have plowed old ground for some readers, and we hope we have stimulated some new thought processes for others. If it seems a bit lengthy, we can only observe that management evaluations are not accomplished with simple one-liners. Managing is a very broad and complex occupation. Even in the area of safety, we have seen that there are many facets to managing it.

At the end of this chapter, we have given our version of a Management Evaluation Checklist, in which we have tried to bring together all of the elements, activities, and measurement approaches, into one cohesive step-by-step sequence. We hope that the reader will refer to it at any time while studying this chapter.

Employee Perceptions

A recent article, by Charles W. Bailey and Dan Petersen, in *Professional Safety* magazine (February, 1989) reported on a nine-year study in the railroad industry on employee perceptions of safety in their work situations. It showed a striking correlation between what employees thought of their employers' safety posture, and what that posture actually was. To safety auditors, this can be an extremely valuable assist in their evaluation of management's safety performance. Interestingly, the Bailey/Petersen study also showed that supervisory and management people tended to identify the same areas of strengths and weaknesses that the hourly wage-earners did. We are not quite sure what that means, since the thought occurs—if management is aware of their own deficiencies, why do they still exist?

It is a truism that it doesn't really matter what I do or say to you; what matters is what you think I have said or done. In other words, your perceptions of me influence your interaction with me—not my actual performance. This same concept, according to the theory, applies in employee/employeer relationships. In the sphere of safety, it is what the employee thinks management's safety stance is that really counts and influences the employee's own approach to safe behavior.

We have used employee perception surveys for a number of years as a tool to measure the culture of an organization. We feel that they provide valuable clues for auditors, but they must be careful to distinguish between the two quite different kinds of clue that they can receive. Employee perception surveys may indicate true deficiencies, *or* it may indicate a failure to communicate reality on the part of management. We do not believe that survey results should be accepted on blind faith, but rather they should be used as pointers —indicators of areas that need deeper probing to reach a valid assessment.

We have also used perception surveys on several occasions when management asked for help in understanding why their accident and injury experience seemed to be out of control. Here are situations where problems clearly exist, but the reasons could not be seen in the course of everyday activities. These were also, by the way, cases of very proactive management, using safety auditors as a resource. One perception survey found that average employees did not think their supervisors and managers considered their safety foremost. The consensus was that if they could get a job done more quickly by taking chances, omitting safety precautions, or crutching faulty pieces of equipment, their management was willing to accept that result, and look the other way.

As a result of that perception, which overrode well-documented and extensive training, and repeated exhortations by management to follow the rules and report safety problems, the workers behaved as though their perceptions were true. To them, they were. Management's sincerity was doubted. So shortcuts were taken, problems went unreported and unresolved, and accidents went up.

This kind of finding will be used by proactive managers in positive ways. One such manager became much more visible and interactive in his shop area. He coached his people, shut down balky and unsafe machinery for repairs, and congratulated individuals when he observed a truly safe activity or behavior. Another, unfortunately, took us to task for being so presumptuous as to criticize his management methods and style. What did we know about managing his department? He dismissed our finding and, of course, accidents continued in his operation, because he rewarded fast work, not safe work.

A negative employee perception is therefore an ambiguous finding, but not to be either taken on faith or ignored. It indicates either a true management deficiency in safety philosophy, or a failure by management to communicate and convince their people of where their priorities lie. Either way, proactive managers can deal with such findings and correct the perceptions of their expectations.

We have included in this chapter a sample perception survey questionnaire, one that we have used many times. The statements deal with a number of basic topics, and the reader will note that many of them are similar, or opposites, or redundant. We have scrambled them so that any given general topic does not show as an obvious series. The code letters following each statement (D, F, C, etc.) refer to the general topic of perception being evaluated. We have tabulated our questionnaire in both the topical and scrambled versions (Tables 10.1 and 10.2), so that the reader can see the organization of the statements, as well as a working model.

We do not suggest a mechanical approach to evaluating the re-

Table 10.1. Perception Questionnaire (topical version).

I. Alcohol and Drugs (D)
 1. People use alcohol and drugs in this plant, or come to work under the influence.
 2. Alcohol and/or drugs are not a safety problem in my work area.
 3. The company's drug and alcohol screening program is fair.
 4. The company's drug and alcohol screening program is ineffective.
 5. I am glad that the company takes positive and effective action when drug and alcohol abuse is discovered.
 6. The drug and alcohol program looks good on paper, but nothing is ever done when a problem is found.

II. Communication (C)
 1. It is easy for me to talk with my boss about a safety problem
 2. When changes happen here, we usually don't know why.
 3. If I bring up a safety problem, my boss acts like I'm the one at fault.
 4. Management normally tells us the reasons for things like changes in our operation, why we need to work overtime, and what they are planning.
 5. Management doesn't listen when we have safety matters that they should know about.
 6. My boss thanks me for bringing attention to a safety problem, or for asking questions about the safety of our operation.

III. Employee Involvement (E)
 1. Management really wants to know our ideas about making the operation safer and better.
 2. Nobody ever talks to us or comes out to the plant when they are planning or changing an operation.
 3. Management respects our intelligence and integrity.
 4. Management considers us a necessary evil in getting production out.
 5. They talk to us a lot to get our ideas when new or changed operations are being planned.
 6. Sometimes our procedures don't work because nobody has come out to see whether they do, so we have to figure it out our own way.
 7. In our area, people really do take responsibility for safety.

IV. Management Involvement (M)
 1. Management talks safety, but they don't back us up if an unsafe situation might slow production.
 2. Safe operation really is my manager's first priority.
 3. Top management is two-faced about safety, so my boss is caught in the middle.
 4. Our management understands our work and our problems.
 5. Management is consistent in its priorities, goals, and directives regarding safety.
 6. Management and supervision stop unsafe operations immediately until the problem is fixed.
 7. Management follows the same rules as the rest of us when it comes to safety requirements.

(continued)

Table 10.1. Perception Questionnaire (topical version). *(Continued)*

  8. When an accident happens, management goes looking for someone to blame.
  9. If we have an accident, our management looks for unsafe conditions as well as unsafe acts.

V. Immediate Supervision (S)
  1. My boss only compliments us on getting the job done on time.
  2. My boss has complimented me on safe work practices.
  3. Good safety practices are kind of laughed at in my department.
  4. Safety incentive awards have a continuing positive influence on workers' safety consciousness.
  5. The safety incentive program is unfair and ineffective.
  6. The only time my boss talks to me is if I screw up.
  7. My boss tries to support us and get safety problems corrected, but the people above only push production.
  8. I work with people who do not work safely, but the boss just looks the other way.
  9. My boss says "no shortcuts," then praises someone who gets a job done so fast that they must have taken some.
  10. Discipline for safety violations is handled fairly in our area.

VI. Safety meetings (SM)
  1. We usually talk about important subjects in our safety meetings.
  2. In our safety meetings, the boss does all of the talking.
  3. Most everyone on our crew gets involved with comments and questions during our safety meetings.
  4. Safety meetings in our area are a waste of time.
  5. Our boss really tries to pick appropriate safety meeting topics, and teach us how to do our jobs safely.
  6. It seems that the boss is just as anxious to get our safety meetings over with as we are.

VII. Priorities (P)
  1. Safety can't be our number one priority when we're busy.
  2. Our boss considers safety in all that is done and in every decision that is made.
  3. Safety is our number one priority.
  4. The boss says that we shouldn't take shortcuts, but then doesn't give us enough time to do the job right.
  5. Time allotted for various jobs is usually about right to do the job safely.

VIII. Facilities and Equipment (F)
  1. If something needs safety repair or maintenance, management will do everything possible to get it fixed quickly.
  2. Our equipment is in generally good condition.
  3. We have a lot of equipment breakdowns, so we usually have to figure out some work-around to get by.

Table 10.1. Perception Questionnaire (topical version). *(Continued)*

    4. Our plant and equipment are in disrepair, and that causes some real safety hazards.
    5. It takes too long to get a safety problem corrected.
    6. My boss never lets us start a job until all of the safety requirements are met.
    7. There is no point in bringing up a safety problem; nothing is going to be done about it anyway.

IX. Training (T)
    1. Most people in our department know their jobs well enough to do them safely.
    2. Nobody can remember all of the safety rules that they lay on us.
    3. The safety rules are too strict and too many.
    4. When I came to this department, I mostly had to figure the job out for myself.
    5. I got very good on-the-job training from my boss when I joined this department.
    6. The orientation and classroom training was boring.
    7. My classroom training really told me a lot about the company and made me proud to be a part of it.

sponses that auditors receive from this, or similar, perception questionnaires. Rather, we recommend that the data be digested as a depiction of a culture. Quantitative treatment of the answers, in our opinion, will get one bogged down in trivia, while the true purpose of the survey becomes obscured.

In a department of, for example, one manager, six foremen, and 47 workers, we would suggest asking the manager, foremen, and 15-20 of the workers to complete the questionnaire. The answers would be tabulated as to positive, neutral (don't know), or negative. If there was a negative consensus in any particular category, it would be clear that this area needs attention. The manager's efforts would have a focus. Too many negative consensuses are obviously a strong clue that safety is not important in the departmental culture. Safety auditors should remember that they are finders and reporters of information, and the employee's perceptions are simply another piece of information. It is not the auditor's role to judge, or to correct.

Table 10.2. Perception Questionnaire (working version).

	AGREE	DIS-AGREE	DON'T KNOW
1. Management normally tells us the reasons for things like changes in our operation, why we need to work overtime, and what they are planning. (C)			
2. It takes too long to get a safety problem corrected. (F)			
3. I got very good on-the-job training from my boss when I joined this department. (T)			
4. The company's drug and alcohol screening program is ineffective. (D)			
5. Management considers us a necessary evil in getting production out. (E)			
6. Management is consistent in its priorities, goals, and directives regarding safety. (M)			
7. The safety incentive program is unfair and ineffective. (S)			
8. Our boss really tries to pick appropriate safety meeting topics, and teach us how to do our jobs safely. (SM)			
9. My boss tries to support us and get safety problems corrected, but the people above only push production. (S)			
10. Safety is our number one priority. (P)			
11. People use alcohol and drugs in this plant, or come to work under the influence. (D)			
12. Most people in our department know their jobs well enough to do them safely. (T)			
13. The boss says that we shouldn't take shortcuts, but then doesn't give us enough time to do the job right. (P)			
14. We usually talk about important subjects in our safety meetings. (SM)			
15. If we have an accident, our management looks for unsafe conditions as well as unsafe acts. (M)			
16. Management doesn't listen when we have safety matters that they should know about. (C)			
17. My boss only compliments us on getting the job done on time. (S)			
18. It is easy for me to talk to my boss about a safety problem. (C)			

Table 10.2. Perception Questionnaire (working version). *(Continued)*

	AGREE	DIS-AGREE	DON'T KNOW
19. My classroom training really told me a lot about the company, and made me proud to be a part of it. (T)			
20. Sometimes our procedures don't work because nobody has come out to see whether they do, so we have to figure it out our own way. (E)			
21. The drug and alcohol program looks good on paper, but nothing is ever done when a problem is found. (D)			
22. Top management is two-faced about safety, so my boss is caught in the middle. (M)			
23. Management really wants to know our ideas about making the operation safer and better. (E)			
24. Good safety practices are kind of laughed at in my department. (S)			
25. If something needs safety repair or maintenance, management will do everything possible to get it fixed quickly. (F)			
26. Most everyone on our crew gets involved with comments and questions during our safety meetings. (SM)			
27. We have a lot of equipment breakdowns, so we usually have to figure out some work-around to get by. (F)			
28. The orientation and classroom training was boring. (T)			
29. When changes happen here, we usually don't know why. (C)			
30. Management and supervision stop unsafe operations immediately until the problem is fixed. (M)			
31. My boss thanks me for bringing attention to a safety problem, or asking questions about the safety of our operation. (C)			
32. Management follows the same rules as the rest of us when it comes to safety requirements. (M)			
33. My boss says "no shortcuts," then praises someone who gets a job done so fast that they must have taken some. (S)			
34. Safety can't be our number one priority when we're busy. (P)			

(continued)

Table 10.2. Perception Questionnaire (working version). *(Continued)*

	AGREE	DIS-AGREE	DON'T KNOW
35. The safety rules are too strict and too many. (T)			
36. I am glad that the company takes positive and effective action when drug and alcohol abuse is discovered. (D)			
37. My boss has complimented me on safe work practices. (S)			
38. They talk to us a lot to get our ideas when new or changed operations are being planned. (E)			
39. Our plant and equipment are is disrepair, and that causes some real safety hazards. (F)			
40. When I came to this department, I mostly had to figure the job out for myself. (T)			
41. If I bring up a safety problem, my boss acts like I'm the one at fault. (C)			
42. In our area, people really do take responsibility for safety. (E)			
43. Nobody can remember all of the safety rules that they lay on us. (T)			
44. Our boss considers safety in all that is done and in every decision that is made. (P)			
45. My boss never lets us start a job until all of the safety requirements are met. (F)			
46. Time allotted for various jobs is usually about right to do the job safely. (P)			
47. Safety meetings in our area are a waste of time. (SM)			
48. Discipline for safety violations is handled fairly in our area. (S)			
49. When an accident happens, management goes looking for someone to blame. (M)			
50. Management respects our intelligence and integrity. (E)			
51. Safety incentive awards have a continuing positive influence on workers' safety consciousness. (S)			
52. Our management understands our work and our problems. (M)			
53. There is no point in bringing up a safety problem; nothing is going to be done about it anyway. (F)			

Table 10.2. Perception Questionnaire (working version). *(Continued)*

	AGREE	DIS-AGREE	DON'T KNOW
54. It seems that the boss is just as anxious to get our safety meetings over with as we are. (SM)			
55. I work with people who do not work safely, but the boss just looks the other way. (S)			
56. Safe operation really is my manager's first priority. (M)			
57. Nobody ever talks to us or comes out to the plant when they are planning or changing an operation. (E)			
58. Alcohol and/or drugs are not a safety problem in my work area. (D)			
59. Management talks safety, but they don't back us up if an unsafe condition might slow production. (M)			
60. In our safety meetings, the boss does all of the talking. (SM)			
61. Our equipment is in generally good condition. (F)			
62. The only time that my boss talks to me is if I screw up. (S)			
63. The company's drug and alcohol screening program is fair. (D)			

Management Systems

Management students and practitioners have invented systems for virtually every function that managers perform. Sometimes, they get caught up in their systematization to the point where their systems are managing them, instead of the reverse. We have seen managers who are so devoted to their time-management notebooks that they don't interact with other people; they are constantly jotting down notes and cross-referencing entries, almost ignoring the people that they are supposed to be conversing with. Others are so beholden to their cost performance reports, or their production schedules, that reasonable discussion of problems or changes is virtually excluded.

Management systems are of value to the extent that they efficiently assist managers in achieving their operating objectives. If systems are overly cumbersome, or they take up too much of the managers' time to use, or if they really do not promote the achievement of the managers' goals, then they are useless.

We are only going to consider management safety systems in this section, all the while recalling our premise in an earlier chapter that truly dedicated and proactive managers never separate safety from the rest of their job. The safety systems that they have in place, and the degree to which they employ and enforce them, are a very good measure of the managers' safety performance.

In our frame of reference, safety systems are the built-in procedures and controls by which the safeness of operations and tasks is established and maintained. This would encompass work disciplines such as hot work permits, confined space entry procedure, employee safety training and hazards communication, operations planning reviews, lockout procedures, and preventive maintenance schedules. Many of these systems, if not all, may have been suggested or provided by the area safety professionals in their role as a management resource. The measure of operating managements ownership of the systems is how they have adopted and used them.

In the Management Evaluation Checklist at the end of this chapter, we have included a number of entries designed to measure the adequacy of these safety systems. We think that they are rather self-explanatory, so we won't reiterate them here. The point that we wish to emphasize now is that safety auditors should ask for, and confirm, evidence that the systems are used and followed when applicable. In other words, look beyond the policy statement and procedure, and find out whether it is done that way. If the managers can produce documented instances of compliance, such as a file of hot work permits, or a file of preventive maintenance worksheets, etc., the auditors can comfortably give a positive evaluation of that item. However, an inability to document system discipline, while not incontrovertible evidence of noncompliance, is at least evidence of a need for managers to organize their record-keeping.

Delegation

We have, throughout this book, talked about, defined, measured, and graded management's safety responsibilities and accountabilities — their safety performance. We may have tended to give the impression that operating managers are supposed to be full-time safety engineers. It certainly would not be a liability to them or their operation if they sometimes behave as though that were true. However, lest we lose sight of reality, we think that it might be appropriate here to address the idea of a delegation of safety oversight. Conceptually, we have no more objection to managers delegating safety matters to assistants than we do to their delegating machine operation and repair, or product inspection and testing, or waste disposal. It is actually an essential element of successful management that the manager delegate many, usually most, of the chores of running an organization to qualified and competent subordinates.

In that context, we then want to be able to distinguish, in an observable and objective way, delegation as distinct from abdication. Delegation never absolves. Operating managers cannot be relieved of accountability by the laxity or failures of their subordinates. Delegation also cannot be interpreted to mean total inactivity on a matter. Delegation simply means turning over the routine, the repetitive, the less creative elements of the work to assistants who know what needs doing, and are assigned time and authority to do it.

For example, a manager may hold a production planning meeting every Friday afternoon, to plan and prepare for the following week's activities. He or she will, hopefully, review the status of crews, equipment, supplies and materials, and open items from previous such meetings—including items pertaining to safety. However, the manager very well might not personally record minutes or notes. Also, he or she probably has not personally retrieved last week's notes with open items and assignments. The Manager will have delegated those chores to an administrative assistant, or perhaps one of the foremen or operators. He or she will expect that subordinate to come to the present meeting prepared to review last week's assignments and record this week's activities.

Another example: in last week's meeting it came to light that the main circulating pump for a safety critical cooling system was down for repairs, and that the cooler was operating on the backup pump. The manager gave orders that a portable pump be brought in and put in stand-by readiness while the main pump was repaired. All of this was duly noted by the designated "scribe," and brought up at this week's meeting by this subordinate. The new status was reported by the repair people, the stand-by pump confirmed ready, the expected date for availability of the main pump estimated, etc. The "scribe" had all of that jotted down in this week's new notes. This is effective delegation in action.

Record keeping is certainly not the only task that proactive managers can delegate in safety. Safety training and hazards communication, proficiency and qualification evaluations, engineering studies of the safety of the equipment, daily work readiness checklist completions, and recycle and inspection of personal protective equipment are just a few of the system implementation routine tasks that can and probably should be delegated.

Managers who delegate appropriately will never lose the visibility necessary to assess the adequacy of the implementation. They don't have to train every one of their operators in tank entry procedures, in order to receive the feedback that each one is trained. Their in-place systems will provide the documented feedback. These managers will see to it that their current systems do provide traceability and feedback, or they will devise ones that will.

Passive managers who delegate and forget about it are really trying to sidestep their responsibilities. They may succeed for a time in absolving themselves of any safety responsibility, particularly if they have equally passive bosses. Auditors are not very likely to find effective management safety systems in use, even if they do exist letter-perfect in the policy manual.

The sign of effective safety delegation is in the policies of managers who have the needed information, statuses, on-going action assignments, and due dates at their fingertips. The sign of managers who have integrated safety into their operations management style is when they use the information that is at their fingertips to make decisions and direct their operations.

Now for that promised Management Evaluation Checklist that we have referred to several times in this chapter.

Discussion of a Management Evaluation Checklist

We have arrived at the point of undertaking that self-appointed and presumptuous task of evaluating management from the standpoint of their safety performance. We have expounded at some length on what identifies their involvement and commitment, what activities they should have on-going, the value and meaning of employee perceptions, the safety systems or practices that they should have in place, and the value of effective delegation that provides the necessary feedback for proactive managing. Now we need to put all that in terms that the Safety Auditor can follow on the floor and in the office, to arrive at a reportable evaluation.

The premise that we started with was that we cannot judge by, speculate based on, or impose our own standards. The checklist that follows later in this section covers a lot of territory, and auditors must be cautious not to pursue areas and items that are not part of the acknowledged standards of the organization. It is embarrassing, in any facet of auditing, to report a finding for which there is no requirement. This is particularly true of management evaluations.

It will be recalled that our advice earlier in this work was that safety auditors should promulgate standards for top management review and eventual adoption. If that effort was not undertaken, or is not completed, or has already failed, we recommend not doing management evaluation audits until some standards have been adopted. In that situation, the checklist below can be used as the basis for proposed standards, based as it is on the descriptions of activities and systems that we have already discussed.

Management Evaluation Checklist

I. Involvement/Commitment.
   1. Was there any controversy over plans to accompany the manager during daily activities?   Yes   No
   2. If the manager objected, describe the reasons for this. _____
   3. Does the manager conduct periodic production planning meetings?   Yes   No
   4. How many items of discussion were covered? _____
   5. In how many of these were safety matters raised or discussed? _____
   6. How many others were there where one or more safety considerations should have been discussed and were not? _____
   7. Describe the item(s) and why safety should have been brought into the discussion (use a separate sheet).
   8. How many other group or individual meetings or conversations were held in your presence that concerned the managers operations? _____

(continued)

Management Evaluation Checklist *(Continued)*

9. In how many of those that concerned operations was safety addressed or considered? _____
10. In how many should safety have been discussed and was not? _____
11. Describe the topic of discussion and why safety should have been a part of the conversation (use a separate sheet).
12. How many times did the manager go into the operating area to see how operations were proceeding? _____
13. How many times did the manager talk to operating people about the safety of their individual jobs or work situations? _____
14. How many positive, complimentary, or constructive comments were made? _____
15. How many derogatory or negative comments or criticisms were made? _____
16. How many unilateral operating decisions did the manager make that you are aware of? _____
17. In how many was safety a consideration? _____
18. In how many others should safety have been considered and was not? _____
19. Describe the nature of the item(s) at issue, and why safety should have been a consideration (use separate sheet).
20. How many total operating items did the manager work on in the course of your observation (total of questions 4 + 8 + 16)? _____
21. How many of these were issues where safety should have been part of the issue? (total of questions 5 + 6 + 9 + 10 + 10 + 17 + 18)? _____
22. In how many was safety actually considered before a decision was made (total of questions 5 + 9 + 17)?
23. What is the percentage of times that safety was considered when it should have been (number in question 22 divided by number in question 21 times 100)? _____

II. Activities
1. When queried, did the manager provide a written or verbal list of his primary operating hazards?   Yes   No
2. Could the manager identify how those hazards are controlled to his or her own satisfaction?   Yes   No
3. Could the manager explain what factors made these controls acceptable?   Yes   No
4. Could the manager rank the hazards in terms of their seriousness?   Yes   No

Management Evaluation Checklist *(Continued)*

5. Is there evidence that the manager and/or the supervisory people have reviewed and approved (concurred with) operating procedures and equipment designs and layouts?   Yes   No
6. How many procedures, plans, designs, and layouts did you look at relative to this operation? _____
7. How many of those were concurred or approved by the manager or the manager's designee? _____
8. What documented evidence did the manager provide regarding operational safety reviews, and the correction of deficiencies? _____
9. How many safety deficiencies did the manager identify? _____
10. How many of these were still uncorrected? _____
11. Does the manager have an MBWA plan?   Yes   No
12. How many times a week does the plan call for an MBWA tour? _____
13. Did the manager conduct the tours as planned?   Yes   No
14. What was the manager's percentage of completions? (number in question 13 divided by number in question 12 times 100) _____
15. Does the manager make notes and keep a record of MBWA activity?   Yes   No
16. Do the open items from the managers' MBWA observations reappear as items of discussion in the planning meetings?   Yes   No
17. Does the manager's schedule for MBWA cover all areas of the operation?   Yes   No
18. Is there evidence that the manager looks for root causes of the deficiencies noted in the MBWA tours?   Yes   No
19. What evidence does the manager have of communication to upper management of safety problems? _____
20. During MBWA, when a safety problem was identified, did the manager explain to the person in charge what the problem is, and why it is a problem? Did the manager coach that person regarding what was expected? Describe examples. _____
21. Was the coaching done in a positive way?   Yes   No
22. Does the manager have a list or log of observations for each employee (or crew), with some positive comments?   Yes   No
23. Does the manager have documentation of investigations of injuries and accidents in the operating area?   Yes   No

(continued)

Management Evaluation Checklist *(Continued)*

24. Is there evidence that the concluded cause(s) have been corrected? Yes No
25. Does the manager have a comprehensive tracking system to help keep track of questions, open items, and action assignments? Yes No
26. How many times did the manager use it in the course of this audit? _____ times in _____ days.
27. Describe its use and the manager's actions when an item had not been completed. _____

III. Employee Perceptions
   1. Do employees believe that there is a drug and/or alcohol use problem in this department? Yes No
   2. Do they approve of the company's policy and way of dealing with it? Yes No
   3. Do the employees feel comfortable talking with the manager about a safety problem? Yes No
   4. Do they feel that they are included and consulted about what is happening or being planned? Yes No
   5. Do they perceive that safety is a part of the company's or department's culture? Yes No
   6. Is the manager, and/or upper management, seen as sincere about correcting and improving safety problems? Yes No
   7. Do they feel that safe performance is adequately rewarded by supervision? Yes No
   8. Is supervision and management seen as positive in their dealings with the workers? Yes No
   9. Is there a perceived difference in priorities between first level supervision and management? Yes No
   10. Is the content of safety meetings considered meaningful to the employees' work situation? Yes No
   11. Do they feel that some matters drive decisions at the risk of their own or others' safety? Yes No
   12. Do they think that the manager "looks the other way" when risks are taken, in the interest of getting the job done ahead of schedule or at less cost? Yes No
   13. Is there a feeling that corrective actions to safety problems are too long in coming? Yes No
   14. Do they think that the equipment and machinery they work with is safe to use? Yes No

Management Evaluation Checklist *(Continued)*

15. Do the employees feel that they and their co-workers have been trained to do their jobs safely?   Yes   No
16. Do they think that the safety rules are sufficient and appropriate for the job?   Yes   No

IV. Management Systems
1. Does the manager have a written training plan for each type of job category?   Yes   No
2. Does the plant or department train personnel in the proper use of needed personal protective equipment?   Yes   No
3. Does the area have written instructions as to the need and criteria for PPE maintenance/replacement?   Yes   No
4. Is there a written policy and training plan for employee hazard communication?   Yes   No
5. Is there a record kept of who has received area-specific training in job and area hazards?   Yes   No
6. Is there a mechanism in place that labels all hazardous materials that enter the operating area?   Yes   No
7. Is there a mechanism in place for inspection of the area or building prior to beginning maintenance or construction activities?   Yes   No
8. Does the mechanism evaluate the safety of using welding equipment, open flames, or other heat-producing items (hot work permit)?   Yes   No
9. Is there a written lockout procedure (zero energy state) for repairing/maintaining machinery?   Yes   No
10. Is there a formal procedure for entering confined spaces such as tanks, tunnels, ovens, pits, etc.?   Yes   No
11. Does the procedure provide for the testing of the atmosphere and criteria for entry?   Yes   No
12. Is there a policy requiring plans, procedures, designs, and layouts to be formally approved and accepted by operating management and other disciplines?   Yes   No
13. Is there a written requirement for periodic monitoring of hazardous locations or materials?   Yes   No
14. Is there documentation of exposure measurement and compliance with the requirement?   Yes   No
15. Is there a formal preventive maintenance schedule, that is coordinated with production operations planning?   Yes   No
16. Are there records of having kept to that schedule?   Yes   No
17. Is there a policy or directive requiring, and describing the method of

(continued)

Management Evaluation Checklist *(Continued)*

   implementing, periodic housekeeping, cleanup, and organizing of the work area? Yes No
18. Are there policies and procedures that spell out how waste materials, scrap, and other items for discard are to be handled? Yes No
19. Is there a procedure in place controlling the use of, accumulation of, and precautions for flammable and combustible materials? Yes No
20. Is there a procedure for protecting, and controlling the activities of, visitors and guests in the operating area? Yes No
21. Are visitors and guests advised of the hazards, escorted, and/or required to wear PPE if entering a hazardous area? Yes No
22. Are there conspicuous and informative signs and warnings advising of hazards and needed precautions or procedures for entry to the operating area? Yes No
23. Are pipes, valves, switches, emergency equipment, and energized equipment properly and conspicuously labeled? Yes No
24. Is there a written emergency/disaster response and recovery plan? Yes No
25. Is there a loss evaluation, analysis, and control methodology in place for prioritizing efforts and resources? Yes No

V. Delegation
1. Does the manager keep a record of the tasks that have been delegated to subordinates, to whom and when results or answers are due? Yes No
2. Does the manager follow up on these tasks in the regular planning or staff meetings? Yes No
3. Describe the manager's way of dealing with subordinates who report incompletions or inabilities to complete a delegated task. _____

_____

4. Are the delegated tasks sufficiently detailed and unambiguous as to be readily understood? Yes No
5. Is the manager willing to reassign a task to a higher level or outside the department when it becomes clear that greater authority or talent is needed? Yes No
6. Does the manager make an effort to verify the correctness and completeness of task completions? Yes No
7. Does the manager have a current status available on all delegated safety assignments? Yes No
8. Is there evidence that the manager uses the current status data in making operating decisions? Yes No

How To Use the Checklist

As can be seen now from a brief glance at the preceding section, the checklist is organized in a manner that parallels the topical sections of this chapter. Each Roman numbered part of the checklist corresponds to a section. So the first twenty-three questions deal with Management Involvement and Commitment, the next twenty-seven with Management Activities, and so on.

We would endeavor now to guide auditors through the checklist, and discuss some approaches and ideas on how to use it. The first two questions of Part I deal with the issue of how the auditors were received, and should be quite straightforward in completing. A simple factual summary of the reasons, or lack of, will suffice, if there was any controversy or resentment. As we have said before, management evaluation audits should not be undertaken unless and until the management performance standards and protocols for evaluation are established in policy and approved by *TOP* management. Given this, the operating management's reaction to the audits and auditors is simply a finding, either positive or negative.

A simple count of topics discussed in the managers' own or other meetings and discussions, and a simple count of the number of times that safety entered the conversation, answers the fourth, fifth, eighth, ninth, twelfth, thirteenth, fourteenth, fifteenth, sixteenth, and seventeenth questions. More difficult to ascertain are those questions that ask how many subjects were addressed where safety should have been considered and was not. In order for these questions to be answered objectively, the auditors need to either know enough about the operations and issues to make their own professional assessments, or interview the managers to find out what their reasons were for not addressing safety during the discussion of a particular subject.

It will be relatively rare for safety auditors to be so totally knowledgeable about the technical and engineering aspects as to be able to comfortably "second-guess" the operating people. The auditors should not hold back if they do have intimate knowledge of the technology, but the understanding we espoused in the chapters on planning and preparation is not intended to prepare auditors for replacing the operating managers, or equalling their presumed knowledge of the operation.

Even when auditors have sufficient expertise to reach their own conclusions on whether safety is a factor in a particular topic of discussion, we recommend that they discuss the matter(s) with the managers, to get their perspective on the item(s). They are, after all, the ones being audited, and it is their actions and methods that are to be evaluated — not the auditors'. So we propose that questions 6, 10, and 18 be answered through conversation with the managers themselves.

To get objective answers, auditors should not challenge the managers' handling of the subject. Rather, they should phrase the interview in the context of learning how the managers considered the issue, and what factors they weighed in reaching whatever decision they made. "Could you tell me what were the determining factors in deciding to _____?" will yield more insight into their style and priorities than "Why didn't you find out if the equipment is safe before deciding to _____?"

Questions 7, 11, and 19 are essay types, where the auditors should compile their notes from the interviews. They serve as the basis for finally answering questions 6, 10, and 18. The information the auditors derive from the interviews must be enough to conclude positively that safety was not considered in a matter where it should have been, before it is counted in answering these questions. If the result is inconclusive, and/or the auditors have doubts, it should remain indeterminate, and go on to the next one. One clue that is useful, though not infallible, is that if the subject under discussion by managers is administrative, it is less likely to be one where safety is a factor, than if the subject is operational. For example, merit increases for their foremen, or checking their supply cost billings, are far less likely to carry safety implications than scheduling overtime or reviewing the maintenance schedule.

Questions 20 through 23 are simply summary arithmetic and wholly optional. Some auditors prefer to let facts and observations speak for themselves, while others may wish to give a grade. We suggest that the auditors decide for themselves which is appropriate for the particular situation.

Part II of the management checklist, dealing with observable and documented activities, can be completed almost in its entirety by accompanying the managers for several days, and by reviewing their documentation. The first four questions, however, are interview questions, and necessitate phrasing probing questions and assessing the spontaneity and knowledge that accompanies the responses. The important evaluation point, though, and the reason that the checklist questions are yes or no type, is to simply determine whether they know the significant hazards, and understand and accept the level of hazard control then present.

Questions 5 through 19 are quite straightforward, answerable from the record (or lack of one), and from plant documentation. We won't belabor them further. questions 20–22 are answerable from having accompanied the managers on their plant walk-arounds—the MBWA. Question 20 asks for essay-type observations and the auditors' views on the managers' approach to employee interchanges. It is rather obvious whether they are being critical or instructive, whether they ask and listen and take notes, and if they do, whether they have in their notes a broad cross section of the workers, not just a few.

Questions 23 and 24 concern the managers' involvement in accident and injury investigations, and can best be assessed from their own records. It will probably take a floor check or a planning review to determine whether cause factors have been corrected (question 24). The last three questions are summary questions to determine whether the managers have an inclusive method of keeping up with problems, solutions, and corrections, regardless of how they come to their attention. Whether the safety issues emanate from a staff meeting, an MBWA tour, or an accident, it should funnel into a system that they use to plan, track, and close out such matters. Observation of the managers in action will lead to objective answers to these questions.

Part III of the checklist evaluates operating managers in terms of their employees' perceptions. We provided a perception survey questionnaire earlier in this chapter, and the sixteen checklist items can be answered from the consensus of those surveys. Here, we will provide guidance on how to turn a pile of questionnaires into the answers for the checklist.

First of all, the working version questionnaires must be unscrambled into the topical version. This task can become tedious for those who don't especially thrive on manual detail. It can be a fairly simple conversion task though, if auditors wish to develop a small computer program where they can enter the responses from the working version and have the program give them a tabulation in the order of the topical version.

It is important to remember that the results of a perception survey are just that—perceptions, and not facts. Unless there is a clear consensus on an issue, it is usually not a finding. If there is not at least 60 percent majority answering an item in one way, there is probably no perception finding to report. Also, some of the areas probed in the perception questionnaire are represented by the same statement made in reverse, so that the consensus has to hold in both directions. For example, the first two items in Table 10.1 assess employees' perceptions about whether drug and alcohol abuse is a problem in their work area. They are expressed as opposite views, and they appear in widely separated portions of the working version in Table 10.2. Their answers provide the answer to Checklist Part III, Question 1, which asks whether the employees believe that a drug and alcohol problem exists in their department. For auditors to conclude that there is a perceived drug and alcohol problem, at least 60 percent of the employees should have agreed with the first statement, and at least 60 percent should have disagreed with the second one. The message to management is then quite clear—the perception indicates a real need for management action. It makes no judgement as to what that action should be, whether it is educational, disciplinary, screening, etc. It is, as are all audit findings, information for management to use, or not use, as they see fit, to their best advantage.

On the other hand, it may be significant for management to know what perceptions exist even in a minority of their employees. If only 10 percent feel that safety meetings are a waste of time, for example, a proactive management may decide to improve the quality, or vary the content of the meeting topics. We suggest that a careful review be made of the nonconsensus answers to the questionnaire for meaningful counterpoints.

The second question in Part III of the checklist asked whether the employees approved of the company's policy and practice on drug and alcohol abuse, and can be answered in a manner similar to items I.3 through I.6 of Table 10.1. Question 3 of the checklist follows from whatever consensus comes out of Section II of the perception survey, having to do with communication with management. Again, a consistent consensus is the criterion for a YES or NO conclusion, since the question is a summary question for all of Section II. Question 4 is answered from the results of Section III of the questionnaire on the employees' own involvement in the safety of their department. In consecutive order, questions 5, 6, and 8 of the checklist are answered from Section IV of the survey, 7 and 9 from Section V, 10 from Section VI, 11 and 12 from Section VII, 13 and 14 from Section VIII, and 15 and 16 from Section IX. For any of these checklist questions where a clear-cut consensus is not apparent, the question should not be answered YES or NO, but rather evaluated for salient points that can, together, suggest to management a course of action to enhance employees' perceptions.

Part IV of the checklist addresses management systems. It covers training for employees, personal protective equipment, hazard communication, work permitting, lockout procedures, confined space procedures, planning reviews, plant inspections, exposure monitoring and controls, preventive maintenance, housekeeping, waste management, fire prevention and control, visitor control, labeling and identification of equipment and hazards, emergency or contingency planning, and management initiatives for loss control. The twenty-five questions listed in this section are answerable from one of two sources. Those that query for the existence of a procedure, instruction, record, or plan are answerable by virtue of the managers producing the procedure, instruction, record, or plan. Only questions 6 and 21 through 23, need be assessed on the floor—these having to do with hazardous material labeling, informing visitors of hazards, escorts and personal protective equipment requirements, physical warnings and advisories in hazards areas, and the labeling or identification of equipment.

The auditors are assessing the existence, and in some cases, a specific element of, the several management safety systems that enlightened managers of the 1990s should have in place. Their effectiveness will be measured in the next chapter, on evaluating facilities and equipment.

Finally, the last eight questions, Part V of the Management Checklist, evaluate managers' delegation disciplines and follow-up methods. We do not suggest that there is a standard, in terms of complexity or frequency rate of delegation, against which managers should be measured. If one manager delegates virtually nothing, is not overwhelmed by detail, and gets things accomplished in an effective way, we have no criticism of that capability or situation. If another delegates most detailed tasks, but has a system in place and working, that keeps the manager abreast and informed of things needed to function effectively, we have no quarrel with that either. The questions in this section of the checklist can be answered by documentation or observation. Documentation that the managers should have will show their record of delegated tasks (1), the level of detail and clarity (4), and the status of each (7). The auditors' observations of the managers in action will provide evidence of how and whether they follow up (2), how they deal with subordinates who have assigned tasks (3), what they do about tasks that are floundering (5), verification of completions (6), and their use of statuses in decision making (8). This evaluation of delegation does not assess whether they delegate too much or not enough, but rather how they delegate and follows up, if and when they do delegate.

Chapter 11

Evaluating Equipment and Facilities

Here, we have reached what some readers may have anticipated as the real crux of a treatise on safety auditing. While we do not share that prime billing ranking, we do acknowledge that auditing the physical plant is an essential and central element of thorough safety audits.

Safety auditing is a multifaceted, observational and factual discipline. When evaluating facilities and equipment, the focus tends to narrow to the factual and observable. Yet there is room in a thorough physical audit for professional opinion and analysis. The thrust of this chapter, however, is how to evaluate the visible.

We will talk about the esoteric values of first impressions, and the mundane use of checklists. We will offer some advice on taking notes in the operating area, and reiterate the value of comparing a finding to a true requirement—or, identifying the finding as a professional "druther."

First Impressions

We have finally arrived on the floor of the operation to be audited. All of our planning, the development of maximum credible events, the likelihood and severity evaluations, the sizing up of the operating management, and their effectiveness, are behind, and we enter the arena with (or without) our host. We look across the floor, or into the warehouse, or up into the mountainous scaffolding and platforms of a process plant, and, if we are human, we have an instant impression. A first impression is impossible to avoid.

Is it important? Yes, no, or maybe. The first impression is not likely to be forgotten, but it is subject to confirmation or to refutation. It colors our subsequent evaluation of the details of our audit, unless we are able to put that first impression into its proper perspective. A spotless, everything in its place, freshly painted and orderly plant or area will make a very favorable first impression. Yet there may be insidious hazards lurking there, hazards that will only later come to light. Alternatively, a seemingly chaotic initial appearance may prove to be, ergonomically and safety-wise, an ingenious arrangement of materials, tools and equipment, and people. Either of these extremes may prove to be, technically and systemically, either very safe and compliant workplaces, or laden with accidents waiting to happen.

The first impression is absolutely not conclusive. That is important to keep in mind as the ensuing days unfold. While unforgettable, it is subjective, and objectivity is the name of the safety auditing game. Orderliness and neatness may be the signs of a sick mind, as we paraphrase a truism of the past. However, they may also be the signature of a manager who puts safety into all that is done on the floor. Apparent chaos or (in a less accusative context) turmoil, to the uninitiated, might very well illustrate a brilliant perception of what it takes to perform the mission or operation successfully, which, as we have noted earlier in this book, necessarily includes safely (remember? successfully means safely, intentionally or not).

Nevertheless, any safety audit must begin with the unavoidable first impression etched in the safety auditors' subconscious mind, and they must bring forces to bear on the need to eradicate those first impressions when they begin and continue the audit of the operating facility. We suggest that a very favorable first impression be considered a challenge to auditors to find hidden problems, deficiencies, and shortcomings. Conversely, we propose that a very negative first impression be taken by safety auditors as a bet that they cannot find some very positive qualities in the safety system and/or awareness in that department. In other words, whatever the first impression is, assume it is wrong, and set out to prove your hypothesis. This will tend to open the mind to both positive and negative findings, and result in a much more objective and balanced view of "what is."

If, on the other hand, we experience a rather negative first impression of an operating area, and allow it to fester and grow in our subjective view of the operation, it is inevitable that we will hardly ever find anything good about the safety measures in place there. Similarly, if we see an industrial paradise in our first impressions, and allow that to take root in our minds, we are quite unlikely to bring all of the objectivity demanded to the audit itself. All of this is simply reiterating the age-old truth that objective observers must rely on facts, make their judgments from facts, and not be persuaded nor dissuaded by subjective hyperbole, nor emotional euphoria, in calling things as they are.

How do safety auditors overcome the subjective seduction of first impressions? To a large degree, it is a measure of their professionalism, whether they can don the mantle of realism, or remain captive to the subjectivity of emotionalism. They must focus on the real hazards that the first impression implies. Are there any? Are they serious? Are they emotionally perceived hazards, or are they supportable in fact? Conversely, does the seemingly idealistic scene truly cope with the hazards that have come to be recognized as inherent in the operation? Does a neat and orderly operating area hide truly significant and latent hazards?

A good looking operating area can indeed hide safety hazards that only an in-depth safety audit might bring to light. Our word of caution to novice or veteran safety auditors is to allow the first impression to fade into oblivion, and then, and only then, begin the real audit of the operating area. Let the facts of "what is," contrasted with "what should be" speak for themselves, without the distortion that the emotional reactions of first impressions might generate.

It is offered almost as an irrelevant aside that one person's chaos is another person's dynamism. Similarly, we might observe that one person's stability is another person's status quo, or static doldrum. Everything is relative and we need to keep these perspectives in mind as we deal with operating management. We have constructed a trite chart to remind us all of the relative value of any given situation, as follows:

$$\text{dynamic} = \text{or} \neq \text{static}$$
$$\text{may equal} \quad \text{may equal}$$
$$\text{chaotic} = \text{or} \neq \text{stable}$$

Dynamic may equal chaotic and static may equal stable, in one person's mind, and in another's, chaotic is no substitute for static, and stable is no substitute for dynamic. The situation, as it is in real life, is the one that safety auditors must deal with, evaluate, and report on. Dynamism is not necessarily chaos, and static inertia is not necessarily stability, nor vice versa. Safety auditors must learn to look past the immediate connotation of what is observed to evaluate the observation in terms of its historic significance, as well as its regulatory relevance. All too often, they will find that a certain practice or equipment design was accepted by an authoritative body before, or even without prior knowledge that, some regulatory prohibition existed that proscribed that practice or design. Be that as it may, safety auditors must remain aloof to such "prior politics" and open old wounds. This is where first impressions, reinforced by observation and documented requirements, can be usefully employed for improvement.

Checklists

Checklists are almost as personalized as wallets and haircuts. There are almost as many ways to construct a checklist as there are people who might want to. They can be area-specific, or generic. The former requires more homework prior to the audit; the latter demands more imagination and innovativeness during the audit. Checklists can also be either OSHA-tailored or issue-oriented. The caution of the OSHA-type checklist is that not every generic type of industrial hazard is covered by OSHA, and therefore some vital line of inquiry may be missed completely. An issue-oriented checklist, on the other hand, is the result of having conscientiously studied the operation and identified the real operational hazards involved, then used an existing generic checklist to assess the safety controls in place against them.

There are some audit checklists available that are purported to be comprehensive, but at the same time general enough to fit the generic type, though these require considerable interpretation and imagination to use effectively. Then there are focused checklists, which seek to assess and identify fairly specific hazards, such as injury agencies, or fire sources, or (back to where we started) OSHA violations.

There can be checklists that are for very narrow, but generic, safety issues such as electrical safety, the storage and handling of

EVALUATING EQUIPMENT AND FACILITIES / 171

flammable materials, material handling operations, and hazard communication programs. If this leaves potential safety auditors in a quandary as to how to develop checklists or which one to use, welcome to the club! The answer to this conundrum is in the charter that auditors have received from management, whether they have been directed to implement safety auditing programs, or whether they have opted to propose their own. If safety professionals are proposing to launch safety auditing programs, then the answer must come from their own assessment of what might best fit the circumstances under which they function.

Checklists are not the beginning and the end of safety auditing. They are the hand-held brains of safety auditors, simply because none of us are capable of carrying with us the innumerable roster of rules and regulations under which industry is expected to operate. The most important aspect of safety auditing checklists is that they be comprehensive enough to identify any safety deficiencies that might exist on the operating floor.

Following this discussion are a number of sample checklists that are intended to illustrate the functional usage of checklists in on-the-floor safety audits. It should be noted that, no matter how specific the sample checklists might seem to be, they cannot, and we do not suggest that they can, fit every potential industrial or operational situation. The checklists that follow are offered only as generic examples that can be, at best, adapted by budding safety auditors for application to and in their particular operational environments.

In order to show how the several kinds of checklists can be used, we are going to draw a word picture of an industrial operation with which we are somewhat familiar (at least as it existed a few years back), and then talk the reader through the audit with the various kinds of checklists. Safety auditors can then decide for themselves which one they are most comfortable with.

The operation that we have chosen for this illustration is a commercial printing plant. Its products are labels, wrappers, cartons and boxes, and almost any kind of packaging material that has printed material on it. This plant custom formulates all of its inks to meet the customer's specifications for color, opacity, gloss, and environmental wear, as well as to ensure compatibility with the stock to be printed on. Many of the customers take delivery of their products as flat stock, in the form of pallet-loads of printed sheets. Others want their labels individually finished, so the printed sheet must be cut and recut, and packaged in reams. Others will want continuous rolls, much like the output of a paper mill.

For example, a chewing gum company might order the gum wrapper labels in a roll four feet wide and three thousand feet long. The roll will have a dozen or so wrapper imprints side-by-side across its width, and four or five to the foot through its entire 3000 foot length. Thus a single roll might have $12 \times 4 \times 3000$ or 144,000 wrappers. The gum company will slit and cut each wrapper out of the roll with its own high-speed automated machinery.

A cereal company, however, might want its cereal boxes printed on flat board stock, maybe sixty by ninety-six inches, and containing fifteen to twenty potential boxes. Again, the company will feed the flat sheets into its own highly automated forming, filling, and sealing equipment. Sheet stock products like this would be handled and shipped as palletized stacks perhaps five feet thick and containing two to three thousand sheets, depending on the thickness of the sheet stock.

A third customer, say a canned food producer, may want creamed corn labels individually finished and in thousand-label bundles or stacks. So the printer, after printing the labels on roll or sheet stock, will slit and cut and package individual labels for that customer.

Dozens of other variations can and do arise in this kind of an operation, especially when it has a proactive, service-oriented, and please-the-customer culture. Customers may want more than one kind of label or logo in the same lot. Some will be interested in large packaging boxes, which must be printed one at a time. Printing billboards is always a challenge, trying to create and assemble the various sections, and have them match. However, for our illustrative example, we will let this printing plant be content with three kinds of products—rolls, sheets, and bundles.

A physical facility to produce these products would be something like this. Obviously, there will be the ubiquitous shipping and receiving docks and their attendant product and raw materials warehouses. The product warehouse is full of little other than combustible labels, cartons, and boxes. The raw materials warehouse will have stocks of paper and hard board of course, but hundreds, if not thousands of varnishes, lacquers, oils, polymers, pigments, and other ink constituents. A veritable cornucopia of chemicals, some hazardous, some not so, go into the manufacture of printing inks.

Then there is the ink plant, where each formulation is prepared from batch cards that specify the ingredients, the quantities of each, the order of addition, and the mixing time and any special conditions. Inks can be formulated and mixed in many kinds of mixers, such as the paddle-type, the kneading type, or the vertical blade type. Some continuous-style mixers are used for single product printing operations like newspapers, where little variation occurs from day to day. A standard fixture in any ink plant is the three- or five-roll mill, whose purpose is to grind the pigment particles to as fine a size as can be achieved. The "nip" between the rolls can be adjusted to virtually zero gap, so that as the nonsynchronous opposing rolls pass the raw ink mixture between them, the pigment particles are both crushed and sheared.

Finished ink is often about the consistency of toothpaste or molasses, and in dedicated press operations may be continuously mixed and fed directly to the press by transfer piping. In the custom printing plant, it will probably be batch mixed and handled in tubs or portable tanks for delivery to the press room.

An inescapable aspect of this plant is the forklifts. They load and unload at the docks. They haul the paper and board stock to storage, to conditioning rooms sometimes, and to and from the presses. They carry pallets and drums of ink raw materials to and from the ink plant, and inks to and from the press room. They haul the pallets and rolls of product from one station to the next.

The nerve center of a printing plant of any kind is, of course, the presses. Everything is geared to the presses' schedules. Their schedules are figured out down to the minute. The run time, the cleaning time, maintenance requirements, set-up time, and proofing time are all taken into account. The sequencing of jobs is figured to minimize cleanup time between them, since different kinds of change-overs require different degrees of cleaning. Presses are as variable as any other kind of machinery. Rotary, flatbed, platen, and rubber roll offset come to mind, and we are sure that there are many new kinds that have been developed since we were involved in this area.

A common feature of any press is that it has many synchronized moving parts and pinch points that need guarding. Another is that it requires a steady supply of paper and ink. The material handling operations around the press must continue while the press keeps rolling.

To complete our verbal picture, we must describe the cutting room. For those customers who want it, the sheet or roll stock is taken to the cutting room, where it is fed into one of a variety of machines. Slitters, guillotine cutters, and template cutters are common, all of which can remove a limb or head without a hiccup. Again, material handling around these machines must continue while they slit, cut, and punch.

We hope that this description of the simplified and hypothetical printing plant is adequate enough to give the reader a reasonably vivid mental picture of the operation that we are going to apply our checklists to. Some other features that we might mention, because they have safety implications, are that most, if not all, mixers, mills, and presses are electrically powered. Most forklifts will be electric, also. Some of the raw materials will be toxic by either inhalation, ingestion, or wound entry. Little, if any, crane or hoist handling is used, unless it is a truly huge-scale operation. Many of the thinning and cleaning fluids are highly flammable, as are many of the ingredients themselves. Lighting is questionable, and the noise level around the presses may be excessive.

So let's use a checklist to evaluate the safety posture of one part of this operation. The first one that we propose to apply is a generic one, shown in Table 11.1 as a Safety Audit Checklist. It attempts to cover all fronts, whether they are relevant or not. However, we will apply it to the ink plant in our printing operation. As we described briefly before, the ink plant is centered around mixers and dispersion equipment. A typical mixing room will have a battery of electric mixers of various sizes and types. A weighing and kitting room may adjoin it, where an infinite variety of varnishes and polymers, pigments, and drying agents will be weighed, packaged, and kitted together for a batch of ink. The weighing and kitting room might, in turn, be adjoined by one or more raw materials storage and/or conditioning rooms. In the mixer room, or another adjacent area, there will be ink mills. A surge room or area will store finished ink batches awaiting quality acceptance and delivery to the appropriate press.

The safety auditor will have requested copies of available documentation, such as drawings, layouts, procedures, etc. The checklist

Table 11.1. Safety Audit Checklist — Generic.

The following items are to be evaluated during the course of a safety audit in the operating area. When there are multiple findings in any given entry area, separate notes should be taken.

A. Documentation
   1. Facility Drawings
      1.1 Are there existing drawings or blueprints that show the building and layout as they currently exist?
      1.2 Do these drawings accurately depict the installation as it is, especially including utility (electricity, water, and fuel distribution) lines?
      1.3 Are the equipment locations in accord with the drawing's placement?
   2. Installation Documentation
      2.1 Are there installation drawings that show the as-is condition of the various pieces of equipment and their interfaces?
      2.2 Do the drawings indicate the fire-protection rating of the boundary walls or limits?
      2.3 Is there a list of critical equipment whose status and configuration is considered inviolable?
      2.4 Are the utility requirements for each powered item of equipment specified?
   3. Special Tooling Drawings
      3.1 Do drawings exist that define the technical requirements for special tools and equipment that are essential to the job?
      3.2 Is there a requirement to periodically test energy-producing equipment, such as hoists and cranes, pressurized systems, or radiographic facilities?
      3.3 Has such testing been conducted and documented?
   4. Shop Planning and Procedures
      4.1 Do the planning documents describing the operation exist?
      4.2 Are they descriptive enough to unambiguously describe the process or operation to be performed?
      4.3 Is the procedure/planning document readily understood?
      4.4 Has a Safety designee approved the procedure/planning document?
   5. Work Orders in Progress
      5.1 What outstanding modification/change projects are in progress?
      5.2 What is the intent and effect of each?
      5.3 What is the status of each?
      5.4 Has a Safety designee approved each of the change orders?
      5.5 Can any of these change orders effect the safety of the operation?
      5.6 Describe the effect and evaluate the seriousness of these.

(continued)

Table 11.1. Safety Audit Checklist—Generic. (*Continued*)

6. Other Documentation
    - 6.1 Are there separate documents, or integrated procedures, that prescribe handling methods and equipment?
    - 6.2 Are the MSDSs available on the floor where needed?
    - 6.3 Are there adequate procedures detailing how to operate special equipment and complex tools?
    - 6.4 Is there a copy of the plant safety manual, or an equivalent, available in the work area?
7. Checklists
    - 7.1 Are there operational checklists that itemize the things that should be checked or verified before starting the operation?
    - 7.2 Do these checklists reflect the actual operation, in terms of equipment and processes?
    - 7.3 Are they complete and comprehensive?
    - 7.4 Is there a system for identifying, and tracking through to completion, the noncompliances to the checklists?
    - 7.5 Have identified deficiencies been reported and corrected?
    - 7.6 Do the checklists cover preventive maintenance, operational readiness, equipment, and training?
8. Training
    - 8.1 Is there a document that identifies the training that each person in the department needs in order to perform his or her job correctly?
    - 8.2 Is there an On-the-Job Training (OJT) sequence that prepares employees for the work they are to do?
    - 8.3 Is there a documented record of each employee having accomplished those OJT requirements?
    - 8.4 Does the record show that each employee has met the minimum requirements for the job being performed?
    - 8.5 Do the minimum requirements accurately reflect the skills needed to do the job?
9. Hazards Identification and Analysis
    - 9.1 Are there existing studies that indicate the nature and seriousness of the hazards associated with the operation?
    - 9.2 Are there hazard mitigation recommendations that have not been implemented?
    - 9.3 Is work in progress to do so?
10. Prior Safety Requirements
    - 10.1 Are there previous safety requirements that have been identified by safety professionals?
    - 10.2 Have they been adopted and are they still in effect?
    - 10.3 Is there a history of accident/injury occurrences in this operational area?
    - 10.4 Does it indicate that effective corrective action is being taken?

B. Building and Area
  11. Siting for Explosives

Table 11.1. Safety Audit Checklist — Generic. (*Continued*)

- 11.1 Are there any explosives (OSHA Subpart H) involved in the operation?
- 11.2 If so, is there an approved site plan indicating the intraline, highway, and inhabited building distance requirements for each explosive site?
- 11.3 Does each site comply with both the quantity of explosive and the distance required between sites?
- 11.4 Is there adequate venting for the explosive release expected from an MCE?
- 11.5 Are there adequate fire doors and walls to contain the explosive reaction and prevent propagation?
- 11.6 Are quantity/distance (bay) limits posted for each operating area or building?
- 11.7 Are the limits in accordance with governing regulations?
- 11.8 Is the operation in compliance with the allowable limits?
- 11.9 Are there access warnings, posted personnel limits, and requirements posted at access points?

12. Access/Walkways/Floors
    - 12.1 Does the access to each work station or occupied work area comply with OSHA requirements (OSHA, subpart D)?
    - 12.2 Are they adequately maintained?
    - 12.3 Are there any slip, trip, or fall hazards evident in the work area?
    - 12.4 Are the floors clean, free of litter, and unobstructed?
    - 12.5 Are there adequate railings, cautions, and warning signs to indicate and guard against personnel hazards?

13. Fire Protection
    - 13.1 Are all fire extinguishers inspected regularly and documented as acceptable?
    - 13.2 Are there sufficient fire extinguishers for the area and the operations conducted?
    - 13.3 Is there an installed automatic fire sprinkler or deluge system?
    - 13.4 Has it been tested within the past year?
    - 13.5 Are there posted emergency evacuation plans for employees to follow?
    - 13.6 Is there documentation of training and/or drills in the exercise of the evacuation plan?
    - 13.7 Is the fire protection system endorsed by a recognized fire protection agency?
    - 13.8 Are fire extinguishers and sprinklers/deluges properly located for the hazards that are present?
    - 13.9 Is there an area warning system that informs all personnel in the exposure area of an emergency?
    - 13.10 Are fire extinguishers and other emergency equipment readily accessible (unblocked)?
    - 13.11 Do the fire suppression systems match the fire hazard type?

(*continued*)

Table 11.1. Safety Audit Checklist—Generic. (*Continued*)

14. Electrical
    14.1 Does the electrical system match the class/division of the operational hazards?
    14.2 Is the equipment maintained in visually good condition?
    14.3 Is there a record of inspection/maintenance of the electrical system to ensure code compliance?
    14.4 Are there any "jury-rigged" electrical junctions, cords, or other fixes?
    14.5 Are all breakers, emergency disconnects, switches, and other electrical fittings labeled?
15. Water and Other Utilities
    15.1 Are water, air, vacuum, and other "house" utilities labeled and capacities identified?
    15.2 Are flow directions indicated, when applicable?
    15.3 Is culinary water distinguished from utility water?
    15.4 Is the maximum working pressure indicated on compressed air lines?
16. Lighting
    16.1 Is the inside and/or outside lighting adequate for the occupancy?
    16.2 Are there emergency lights installed as needed?
    16.3 Is there documentation of periodic inspection and maintenance of lighting systems?
17. Ventilation and Hoods
    17.1 Is there adequate ventilation in the operational vicinity of toxic, noxious, or asphyxiant materials?
    17.2 Has the ventilation/evacuation system been designed in accordance with the TLV or PEL of the material(s) involved?
    17.3 Has the ventilation/hood/evacuation system been checked for capacity in the past year?
    17.4 Is there any evidence of deterioration in the system?
    17.5 Are hoods used for other than fume/vapor/gas/dust removal (i.e., storage of materials)?
18. Heating, Cooling, and Air Conditioning Systems
    18.1 Are such systems essential to process efficacy, as defined in the operating instructions?
    18.2 Are they performing according to requirements?
    18.3 Is there evidence of periodic maintenance/inspection?
19. Egress/Escape
    19.1 Are emergency egress paths properly marked?
    19.2 Are all emergency egress routes unblocked?
    19.3 Exterior to buildings and/or shelter, are egress paths maintained free of ice, snow, and other clutter?
    19.4 Are extraordinary means of egress, such as chutes, ladders, "socks," and ramps maintained in usable condition?
    19.5 Do all occupancies have Life Safety Code compliant egress?

Table 11.1. Safety Audit Checklist — Generic. (*Continued*)

20. Storage/Racks/Shelves
    20.1 Is there a designated place for everything used in the operation, such as tools, supplies, solvents, etc.?
    20.2 Are unused items stored in such designated areas?
    20.3 Is such storage neat and orderly?
21. Waste Handling and Disposal
    21.1 Are there procedures for the handling and disposal of process waste?
    21.2 Are they being followed on the floor?
    21.3 Is the pickup and disposal of floor waste in accordance with EPA or state requirements?
    21.4 Are there any outstanding citations or imposed requirements that are not obvious to the casual observer?
22. Personnel Protective Equipment (PPE)
    22.1 Do the operating procedures identify when and under what circumstances PPE must be used?
    22.2 Do they identify specifically what PPE is to be used?
    22.3 Is there documented evidence that the PPE is maintained according to existing requirements?
    22.4 Does the PPE appear to have been so maintained?
    22.5 Are there observations on the floor of violations to the documented requirements?
23. Electrical Installations
    23.1 Is equipment designed for user/functional friendliness?
    23.2 Are all electrical controls labeled and readily accessible?
    23.3 Are safety interlocks appropriate for the operation?
    23.4 Are they maintained intact — not bridged or bypassed?
    23.5 Are all electrical and control components securely mounted and/or installed?
    23.6 Are there out-of-tolerance alarms to warn of unacceptable operating conditions?
    23.7 Does electrical equipment have clearly identified emergency disconnects?
    23.8 Are the disconnects accessible and unobstructed?
    23.9 Is emergency power needed and provided?
    23.10 Are there "temporary" cords or make-shift electrical installations in use?
    23.11 Are any power lines, conduits, or junction boxes in a frayed or disrepaired condition?
24. Mechanical Installations
    24.1 Is the equipment installation "user-friendly," ergonomically designed?
    24.2 Are equipment controls and monitoring devices labeled and informative?
    24.3 Do interlocks and/or lockouts exist to prevent operation under adverse conditions?

(*continued*)

Table 11.1. Safety Audit Checklist — Generic. (*Continued*)

- 24.4 Are load-bearing equipment items (hoists, cranes, forklifts, pallet jacks, stands, etc.) identified with capacity, proof test, and retest data?
- 24.5 Is all process and operating equipment labeled for unambiguous identification?
- 24.6 Does the equipment carry appropriate hazard warnings?
- 24.7 Do pressure/vacuum equipment and systems comply with appropriate codes?

25. Pressure/Vacuum Systems
    - 25.1 Are the control/monitoring systems "user-friendly"?
    - 25.2 Are controls and monitoring items labeled, accessible, and readable?
    - 25.3 Are there fail-safe features, such as relief valves, alarms, trip circuits, etc.?
    - 25.4 Do relief and vent streams discharge to a safe place?
    - 25.5 Are there records of regular maintenance on the equipment and controls?
    - 25.6 Do the records include periodic testing and/or calibration of critical instruments and controls?
    - 25.7 Does the control station prominently show the upper operating limits of the equipment or system?

26. Residual Energy
    - 26.1 Are there mechanical devices, components, or systems that can retain residual energy even when deenergized?
    - 26.2 Are there electrical components that can store energy even when disconnected?
    - 26.3 Are there pressurized or vacuum components that can retain a pressure differential when shut down?
    - 26.4 Are there equipment components that can retain materials in-process after shut down?
    - 26.5 Are procedures and precautions in place to recognize and deal with these forms of residual energy?

27. Workstands
    - 27.1 Are workstands designed for comfort and utility?
    - 27.2 Are load capacities determined and posted?
    - 27.3 Is emergency egress available and in compliance?
    - 27.4 Are there adequate toe-boards, railings, etc.?

28. Material Handling Equipment
    - 28.1 Are forklifts and other mobile material handling units posted with their rated capacities?
    - 28.2 Are cranes, hoists, and jibs posted with their rated capacities?
    - 28.3 Are slings, belts, clevises, eyebolts, and other detachable lifting equipment items posted or labeled with their rated capacities?
    - 28.4 Is there, in place, an operational system to identify or measure the weight of questionable loads to be handled?

Table 11.1. Safety Audit Checklist — Generic. (*Continued*)

- 28.5 Is there, in place, a periodic inspection, maintenance, and testing procedure for material-handling items?
- 28.6 Do records verify compliance with this procedure?
- 28.7 Do operating procedures direct which type of material-handling equipment should be used for each operation?
- 28.8 Is such direction in accordance with regulations and/or good practice?
- 28.9 After observing the use of material-handling equipment, does such equipment have effective controls?
- 28.10 Are the controls labeled?
- 28.11 Are the controls readily accessible to the operator?
- 28.12 Is the emergency disconnect or "kill" switch readily accessible to the operator?
- 28.13 Is there a daily or periodic operator's checklist that is completed prior to the operation of the material-handling equipment?

29. Hand Tools
    - 29.1 Are hand tools used in the process appropriate for the work to be done with them?
    - 29.2 Are they located/stored in such a place that they do not interfere with, nor create a hazard, to the operation itself?
    - 29.3 Are powered hand tools grounded?
    - 29.4 Are all hand tools in good repair?
    - 29.5 Do the operating procedures specify which tools are to be used for each operation?
    - 29.6 Are special tools designed and used for specific purposes?
    - 29.7 Are they considered user-friendly, designed to do the job safely?

30. Flammable Materials
    - 30.1 Are there proper flammable storage cabinets available for the storage of same?
    - 30.2 Do they contain only as much material as they are designed for?
    - 30.3 Are the cabinets properly grounded?
    - 30.4 Are the cabinets properly vented?
    - 30.5 Are there any incompatible materials stored together?
    - 30.6 Is bulk or drum storage in accordance with fire codes, with respect to separation distance, shielding, and lightning protection?
    - 30.7 Are excessive quantities of flammables kept in the operating area (more than are necessary for the job at hand)?

C. Work Practices
  31. Procedures
      - 31.1 Are procedures followed meticulously?
      - 31.2 Do operators stop operations when the procedures cannot be followed, or they recognize an unsafe situation?
      - 31.3 Are procedures referred to when "glitches" arise in the process?

(*continued*)

Table 11.1. Safety Audit Checklist—Generic. (*Continued*)

      31.4 Is there evidence that the procedures are well understood by the operators?
      31.5 Do they use the tools, equipment, and personal protective equipment spelled out in the procedures?
      31.6 Is there evidence that the operating people fully understand the hazards associated with their operation?
      31.7 Do the operating crews maintain a neat and orderly workplace?
      31.8 Where specified, do they request visitors and transients to comply with local safety and/or access rules?
      31.9 Do they maintain ready access to emergency disconnects, fire extinguishers, eyewash, and other emergency equipment?
      31.10 Are materials and products identified with proper labels or other identifying marks?
      31.11 Do the operators make and break grounding connections according to specified sequences?
      31.12 Do they ever perform operations out of procedurally specified sequence?
      31.13 Do they monitor equipment performance, take data readings and record data as specified, in real time?
      31.14 Are there industry consensus work practices that apply to the operation?

D. Miscellaneous
    32. Noise
      32.1 Are data available to document noise levels at various times and situations in the operation?
      32.2 Are the readings within allowable limits?
      32.3 Are adequate hearing protection measures in place?
    33. Vibration
      33.1 Is vibration/oscillation disconcerting at any of the work stations?
      33.2 Are there engineering studies to document the conditions observed as being acceptable?
    34. Fumes/Dusts/Vapors
      34.1 Are such emissions present in the operation?
      34.2 Do procedures specify engineering controls or PPE that must be used to abate the hazards?
      34.3 Are these controls in use in the operation?
    35. Temperature Extremes
      35.1 Are there extenuating environmental exposures?
      35.2 Are there operational controls in place to limit personnel exposure to these conditions?
      35.3 Are they complied with?
      35.4 Are personal protective equipment available for these extreme conditions?

Table 11.1. Safety Audit Checklist—Generic. (*Continued*)

36. Physical/Ergonomic Considerations
    36.1 Are there material-handling steps requiring very heavy or awkward items to be man-handled?
    36.2 Are there situations in which unusual body movements or positions are needed to do the job?
    36.3 Are personnel physically suited to the demands of the job?
37. Communications
    37.1 Are there situations in which a worker is alone while performing his job?
    37.2 Does this worker have a passive communication path with others in case of emergency?
    37.3 Is there a "buddy" system in place for hazardous operations?
38. Confined Spaces
    38.1 Are all confined spaces (pits, tanks, sewers and tunnels, vaults, walk-in chambers, etc.) properly labeled?
    38.2 Are the hazards and access procedures posted?
    38.3 Are the testing requirements for personnel entry included in the area or process procedures?

items should generally be answerable from the documentation available. Obviously, not all questions apply to every operation being audited, so we recommend that one of three basic answers be used—Acceptable, Unacceptable, or Not Applicable. The latter is, of course, used when a question is irrelevant. An alternate terminology can be a simple Yes, No, or N/A. This might carry less of a judgemental connotation.

As applied to the ink plant, the auditor should compare the drawings and equipment descriptions with the actual observed layout and equipment identifications. We are not suggesting using a tape measure to validate the dimensions on the layout, but merely establishing that the relationships and identifications correspond. The questions regarding the content of the installation drawings (2.1–3.3) can be evaluated directly from the drawings.

The checklist items 4.1–4.4 are concluded directly from the review of the shop planning, whether provided before the audit or during it. Ambiguity and specificity are key factors in this element of the evaluation. In our view, management owes it to itself and to its operating people to provide clear, concise, and unambiguous instructions to its operating crews. This not only eliminates the questions about what was intended or planned, but, to the degree that these instructions are unequivocal, relieves management of the onus of not providing plain and understandable procedures, should something go awry in the operation.

For example, in the context of the ink plant, the instructions for the operation of the ink mill should, at the very least, include warnings about the dangers of getting one's fingers or hands caught in the "nip" of opposing rolls. It should also describe the emergency shut-down procedure, the pre-operational checks that should be made before processing a batch of raw ink mix through the mill, and what is to be done if the mill does not check out properly during those pre-operational checks. In addition, the batch-specific instructions should, in one way or another, tell the mill operator what the health hazards are, and what measures should be employed to mitigate them.

Outstanding work orders (known variously as job orders, maintenance orders, "red flag" orders, etc.) are a valuable clue to the proactive bent of the operating management. Given that the order is legitimate, signed, and addressing a particular safety problem (which the safety auditor should validate), the nature of such floor-level initiatives should be assessed by the auditor to judge whether problem corrective actions are indeed responsive to the identified problem. The questions themselves (5.1 — 5.6) are the burden of the safety auditor to evaluate. *If* there are positive corrective actions in process, and *if* they are considered effective, and *if* the safety professional has endorsed them, and *if* there is evidence of diligent prosecution of their completion, then the safety auditor might conclude that a proactive management is indeed trying to upgrade its safety condition.

Section 6 of the checklist in Table 11.1 is a catch-all group intended to find out whether each operator's action, verification, and function has been carefully thought out by those who dictate such operational details. Specific equipment operating procedures, MSDSs, and safety requirements peculiar to a particular operation are considered in light of the MCEs and needs of the operating personnel to operate without loss. This a professional assessment in every sense of the word.

Checklists, the title of this chapter, are pertinent to safe operation, as well as safety auditing. We have alluded to them already. The auditor's questions have only to do with whether such checklists are available, appropriate, used and acted upon in the operation under scrutiny. The questions of Section 7 of Table 11.1 are intended to evaluate this area, and determine whether checklists are a proactive management tool, or simply an exercise that the floor supervisor completes for the record.

Training (Section 8 of Table 11.1) is clearly an evaluation of the attempt by management to provide the necessary understanding, on the part of their employees, as to how to do each one's job with efficiency and safety. The safety auditor is obviously most interested in the latter area—does the training provided give the employees the information needed to do what is expected in a way that does not jeopardize their well-being? These questions are designed to elucidate that central one, and should be answered and evaluated on their own merits. A considerable amount of professional understanding of the operations is necessary to truly assess this element. Knowing the operations, knowing the hazards, and understanding the MCEs that enter into this evaluation (refer to Chapter 5) are essential to this type of evaluation.

In the ink plant, training records should show, as a minimum, what machinery the operating people have to use, what the essential elements of safe operation are, who has been qualified to operate them, and when they were qualified and by whom. Anything less is not an indication of adequate training.

Section 9 of Table 11.1 is the safety auditor's guide to determining how well management has addressed the issue of the hazards extant in the operation. A negative answer to 9.1 is an indication that (1) the question has never been considered, or (2) that the answers have never been considered important enough to obtain. In either case, a reactive management is implicated. If, on the other hand, there are hazard identification and mitigation analyses on hand, pertaining to the operation in question, and positive action was or is being taken to alleviate the problems so identified, then a very positive assessment can be made. The reasons for inaction can also be enlightening, and can form the basis for an assessment of management commitment.

Section 10 of Table 11.1 presumes that there are records of prior safety audits, inspections, or analyses. If there have been none, the line of questioning is moot. It can be viewed as slightly redundant to Section 9, in that the premise is that there have been safety assessments made prior to the present audit. The same management assessment theories apply as in Section 9.

Section 11 is admittedly quite provincial in nature. The reader is prone to ask what relevance explosives safety has to do with the normal industrial operation. The answer is, quite simply, nothing, unless there are explosives involved in the operation, by any stretch of the imagination. To those readers who are in some way involved in explosives operations, it will be readily recognized that this section of the checklist could occupy many more pages. To those readers who have no remote connection with explosives, we suggest forgetting that we included this section.

There is a point to be made here, that will not appear in any regulation or standard that we know of. Many MCEs that a competent auditor or safety professional might identify will look like explosive events, even when there are no bona fide explosives involved in the operation. Operations involving highly flammable gases and liquids, high-pressure equipment, and especially any combination of those two, lend themselves to consideration in terms of explosive events. We would refer the reader to other sources for treatment of these potentialities.

Section 12 of Table 11.1 begins the physical assessment of what is really there on the operating floor. From here through Section 20, floor observations are essential to the evaluation of the various questions posed. We can offer no other advice than to use the checklist during a prolonged tour and inspection of the operating area. Such things as access, egress, working surfaces, in-place fire protection measures and their adequacy, electrical exposures and code requirements, labeling, lighting, ventilation, heating and cooling, and storage can only be assessed on the floor. Sometimes, it is advisable to enlist the expert evaluation or opinion of another party, recognized in his specialty, before assigning a "go/no go" rating in some area such as heating, ventilation and air conditioning, electrical compliance, and/or fire protection/suppression adequacy. Regardless of the assistance that the auditor solicits, the conclusions should be based upon the conditions and practices found in the operating area.

Section 21, on waste management at the operating floor level, is intended to lead the safety auditor into the sphere of environmental affairs, and the determination of whether the control of waste begins at the operating level. It will, at the same time, lend some measure of priority to the more detailed audit of environmental issues that may follow.

In the case of the ink plant, the obvious waste is left-over raw materials — polymers, varnishes, pigments, solvents, and other miscellaneous (though not insignificant) components, and unused ink. Rags and other clean-up materials may also be a part of the overall waste generation and disposal scenario. The safety auditor simply wants to establish factually, at this point, what procedures and practices are in place to collect, remove, and dispose of such materials.

Personal Protective Equipment (PPE) auditing is one of the most important aspects of safety auditing. The underlying assumption in Section 22 of Table 11.1 is that engineering control mechanisms have been exercised to their ultimate, and that the remaining combative measures are protective. Suffice it to say at this point the PPE is a requirement and that the purpose of auditing to Section 22 is to determine how well the PPE requirements are implemented in practice. Questions 22.1 and 22.2 are answerable from the shop planning. Question 22.3 requires research of the records, and assumes that there are such. The last two questions must be evaluated on the floor.

Electrical assessment (Section 23) is a combination of observation, judgment based on observation, engineering analysis, and regulatory compliance assessment. We do not suggest that the safety auditor don the cloak of an expert electrical engineer, and attempt to evaluate or validate the code compliance of an electrical system. We do propose, though, that the safety auditor seek outside assistance in asserting the appropriateness of the electrical system in any given operational installation. The answers to the questions in Section 23 might well come from an indepth interview with the engineer who designed and supervised the construction of the facility. It may come from the evaluation of the system by the operating people themselves. Certainly, a portion of the overall electrical system audit will come from floor observations made by the safety auditor personally. The answers to some questions, such as 23.2 and 23.6, and 23.9 will depend, to a major degree, on the process savvy that the auditor and/or the enlisted assistants bring to the safety audit.

In context of our ink plant, the auditor must have the know-how or the expertise available to determine the "real" requirements of the operating environment, and what electrical codes and advisories apply. If the ink room is subject to occasional releases of flammable vapors, Class I, Division I codes apply. The auditor and/or teammates need to understand both the operational possibilities and the requirements.

Much the same commentary applies to Section 24, on mechanical systems. Floor observations, professional assessment, and expert advice may all be needed to reach a rational determination of the adequacy of these supportive facilities. Testing requirements, and their conscientious prosecution, are offered as major elements in the assessment of the adequacy of an in-place mechanical system evaluation.

Section 25, on pressure and/or vacuum systems, attempts to assess the adequacy of safety controls for the obvious hazards associated with such equipment. Special situations may exist that will require elaboration in the context of those special or unusual circumstances. For example, highly flammable materials under high pressure present a whole new level of hazard that needs evaluation.

Section 26, on residual energy, is a sleeping giant in the safety auditing arena. This relatively obscure phenomenon is a potential killer and wreaker of havoc. Drawing upon our background in explosives and propellants, we would only cite the tremendous difference between the residual energy (and destructiveness) of a quantity of material in a quiescent state, and that same quantity in a reactive mode. The residual energy in a pound of explosive is on the order of a thousand times the potential energy of a pound of an inert mass. The point of this comparison is simply that apparently inert (unenergized) "things" can look very much like the same "things" with a lot of destructive and injurious power. The safety auditor, and the expert assistants, must look for these potentials.

Energized equipment is probably much more subtle than an explosive charge. Lathes, pressure vessels, parked vehicles, and closed circuits are all examples of lurking potential energies. The ink mills and mixers have this potential when all there is to "deenergize" them is an ON—OFF switch. The safety auditor must learn to look past the superficial to the means of truly reducing the residual energy to *zero*.

This implies cut-off switches, or emergency disconnects, for all equipment that is powered, or retains power when deactivated. The residual energy in a deactivated piece of equipment can be quite lethal. There are many documented cases where presses have been turned "off," yet people were hurt when the residual energy in a "turned-off" piece of equipment was released during maintenance or overhaul operations.

The safety auditor is, if so charged by management directives, obliged to understand the possibilities of residual energy in the equipment used in the operation, and assess the potential for dissipation of that residual energy. Questions 26.1 through 26.5 are intended to address those issues. Stored electrical, mechanical, pressurized, and material or chemical energy must be evaluated to determine the control that exists over the residual energy hazard in the operation.

Section 27 of Table 11.1 deals with workstands, and are OSHA-type questions. The first question (27.1) is somewhat subjective, but many OSHA requirements are, too. The assessment of comfort and utility are usually based on the auditor's personal concept of what constitutes ergonomic acceptability. The uninitiated safety professional will have to deal with this issue from an ingenue stance, while the experienced safety professional will bring his previous encounters with OSHA to bear in this evaluation. The best advice that we can offer is to base one's judgment on the precedents of the past, if any, but to use professional judgment as well, if there is no past history upon which to take a stance.

The rest of Section 27 is answerable from an on-the-spot inspection. The existence of adequate capacity designations, emergency egress, and protection from falls or falling objects, are assessible from an on-the-floor audit. The safety auditor should be prepared to observe and be aware of differences between drawings and procedure descriptions and what is seen on the operating floor. There is often a striking distinction between the two. A procedure might very well describe an ideal arrangement of people, materials, and products or components, and the layout drawings might very well show a wholly compliant arrangement of those resources. Nevertheless, the as-is condition may differ significantly from that envisioned ideal. The safety auditor must be in a position to contrast the planned versus the observed, in any of these situations.

Section 28 of the Table 11.1 checklist is concerned with material-handling equipment, and the insurance that it is kept in a functionally safe state of repair. The several questions direct the safety auditor to the floor itself, to the management documents that govern the use and maintenance of handling equipment, and to the records that document compliance with those management directives.

We would insert an editorial, at this point, to the effect that, in our opinion, no safety program is complete without a definitive material-handling equipment maintenance and validation schedule, which includes the proof-testing of that equipment to at least its rated capacity, and, preferably, to some level above the rated capacity. We are well aware of the OSHA minimum requirement for initial proof-testing, and simply take our stance on the question of adequacy, or compliance, versus excellence. An enterprise that plans its strategy around minimum requirements will, in the long run, reap minimum rewards—they might even be negative. Our recommendation is that a management system be in place that automatically schedules and performs 125 percent testing of any and all load-lifting and/or -bearing equipment, at least annually, and that the recording of such testing, and the test results, be in some permanent management record system. The judicious use of such data is obviously within the purview of the particular manager in charge of the operation, and we have discussed these characters at length in Chapters 1 and 2.

Section 29 of Table 11.1 deals with hand tools and their state of repair. It also touches on the question, so critical in the aerospace industry, of whether hand tools are controlled by number and description in the operating area. This aspect is perhaps unimportant in most industries, and the underlying reasoning ignored, where the possibility of "foreign objects" is of no safety concern. It is a "condition of employment" in the explosives and propellents industry that the hazards of foreign objects, such as hand tools, be recognized, and that tool control be exercised to the nth degree.

The question of the adequacy of hand tools extends far beyond the aerospace industry arena, though. The checklist questions of

Section 29 are appropriate to virtually any industrial scene that we can imagine. The answers to the checklist questions 29.1 through 29.7 are obtainable only from an on-the-floor, workplace-to-workplace inspection, or audit. Implied in this section is an on-the-spot inspection of the equipment in use, *and* an assessment of the appropriateness of each, plus a determination of whether these tools are authorized by the official line operating procedures.

Flammable materials handling and control is a discipline and management challenge in any industrial setting that we can envision. The temptation to bring the flammables as close to the operation as possible, in the name of efficiency and cost/manpower optimization, is ubiquitous. Whether the presence of flammables is an extra and unnecessary hazard to the safety of the operation frequently does not enter the equation when management decisions are made in these areas.

In many operations, the issue may very well be considered moot. It probably doesn't matter to those directly involved, whether the acetone and MEK storage cabinet is inside or outside the operating area of a gasoline fractionation unit in a refinery. Yet the presence of flammable solvents in the gasoline area clearly compounds the hazards in that area, and it should be apparent to all that safety will be enhanced if the flammable materials, only tangentially related to the operation itself, are removed from the area. The questions of Section 30 are answerable from the on-the-floor inspection. As is the case with many of the subjects in this portion of the checklist, the underlying issue might be whether the management documentation and/or requirements demand anything better than observed by the safety auditor. Remember always to compare what is with what should be (as defined in management documentation).

The checklist questions of Section 30 of Table 11.1 address only the physical issues that may attend an on-the-floor audit. The open questions, as alluded to above, are whether the quantities are really necessary, whether the necessary precautions are prescribed in the operating procedures or directives, and whether the use of a particular flammable material is essential from a technical or political stance.

The latter question (political expedience) is far more relevant, in many situations, than some would admit. The anathema of *change* is, to some management people, as horrible as a dip in the stock price. The safety auditor should be aware of these potential undercurrents, though the findings or recommendations should never be colored because of them.

This editorial comment is not only germane to the issue of flammable solvents and other materials; it pertains to any fundamental safety matter. A very central question for safety professionals is whether they will stick to their Code of Ethics or become a party to the expedient. Management can make that decision a very crucial one.

The questions in Section 30 are totally answerable from an on-the-floor inspection, and we do not intend to offer any more trite guidance than those checklist items imply. The question of compatibility (30.5) is a possible exception, in that we recommend consultation with the gurus of the MSDSs and/or the chemical wizards of the organization before accepting an unknown material into the operating area.

Part C of Table 11.1 addresses the critical question of whether work practices equate favorably or unfavorably to the expectations of management. The whole premise here is that said management *has* some expectations regarding operating management's control of, and deliverance of, operational performance.

Whether that performance approaches excellence, meets the definition of acceptability (mediocrity), or simply is tolerated, is wholly a management decision.

We will proceed from the assumption that top management does indeed want, promote, and reward excellent performance. The safety professional reading those words might well ask, "if that is the case, why do we need to audit? If excellence is demanded, facilitated, and rewarded, no one will be anything other than excellent." We couldn't agree more! Our message is aimed at the middle management, the operating managers that, presumably, the safety auditor interfaces with while auditing. More often than not, the pronouncements of top management about demanding, and making possible, correct operational performance and practices, and then rewarding it, are unheard by the line manager. Either the message is not received, contrary messages are received, or the message is not believed. In any case, the manager shoots the messenger.

Our message is also aimed at the top management that probably commissioned our auditing services, and who thinks they are facilitators of excellence. If they see that their demands and expectations are not being realized, and are among our readers, they deserve to know why we, at least, think their efforts are failing.

Actual work practices in the plant reflect the integrated sum of all that the workers think it takes to please their bosses, the ones that have jurisdiction over their remaining employed. To the guys who are planning to exit the operation and/or the company, their performance against those perceived requirements for remaining will likely take a nose-dive. By contrast, for the operators who truly want to stay on the job, out of a need for employment, or for the pure joy of it, they will perform exactly as they see their superiors want them to. They will meet their expectations to the letter, as they understand them.

Somewhere in between are the stereotypical employees, who want to keep their jobs, but are only willing to do the minimum necessary to do so. Many managers see their work force in the latter light, benighted as we modern folk might view this perception. It is our premise that, with very few, but acknowledged, exceptions, people want to do the job right. If they fail to do so, we suggest, and with no claim for originality, that it is because they are unable to. They may lack skills and knowledge, they may lack adequate tools and facilities, they may lack leadership and proper priorities. But in general, they lack some vital element that management failed to provide, but should have.

When the opposite case surfaces, that of people who just don't care to put forth the effort to do the job correctly, even when all of the skills, training, and physical resources are available to do it, of course management has little choice but to prune off the offenders. However, we urge that this be the final corrective action, once the premises stated above are validated.

Having integrated another sermon on management styles, let's look into procedural compliance, Section 31 of Table 11.1. The very first of the questions (31.1) is perhaps the true test of a safety auditor's ability to audit work practices. In order to answer this one, the auditor must be in a position of both obscurity and confidence, with respect to the operating people being observed. That, however, is a damn tough assignment. The auditor can only gain a measure of confidentiality by virtue of personality—how convincing the auditor can be that the purpose of the audit is to help effect improvements, not indict the people being observed. The auditor can profess anonymity for the crew members. The auditor can promise to explain, from their vantage point, why something had to be done as it was, due to a lack of some essential element that would have allowed proper execution of the operation. One option is to report, in real time, an observed unsafe practice or deviation from procedures, assert anonymity, and let operating management respond to the finding. This sounds a little bit like an adversarial relationship with the operating management, and we do not mean to promote that kind of antagonism. However, if reticence is encountered in management, this can be used as a tool to reverse it, and to document the need for their action and involvement.

The answer to question number 31.1 will be a long time in formation. It requires lengthy periods of observation, and much one-on-one communication with the operating folks, all with the purpose of finding what prevents total compliance. On occasion, the safety auditor will find that the procedure itself is flawed, and creates a hazard that the operators were smart enough to see, and avoid by deviating from it. In these cases, the auditor reports a hazard and the proper method of alleviating it, which may in fact be the deviation that has been observed. The auditor must also assign action to the procedure writer(s) to obtain their explanation and planned action for alleviating the misdirected procedure.

Questions 31.2 through 31.6 are elaborations of the first question, addressing several specific ways in which procedures may or may not be "meticulously followed." As with question 31.1, floor observation, and interviewing from the vantage point of confidence, are the keys to obtaining answers to them. Questions 31.7 through 31.13 can be evaluated from floor observation, regardless of the degree of confidentiality that the safety auditor has been able to build with the operating people. These are clearly "what is" type observations.

Question 31.14 on consensus standards is not intended to be a ringer; it refers back to the research recommended in Chapter 8, where we talked about advisory standards. ANSI, NFPA, and industry consensuses are all part of the vast body of knowledge that can bear on the safety of a given operation. If and where any of these apply in the operation under audit, this catch-all question is simply intended to remind the auditor that other practices and standards might require the development of additional audit checkpoints or questions. For example, the American Petroleum Institute (API) might recommend that pressure and temperature measuring devices be calibrated to a certain NBS (National Bureau of Standards) level every two months or when repaired. We don't know this to be the case, but if that were the case, and it is far more definitive and restrictive than an OSHA or ANSI standard, it nevertheless becomes a legitimate question for an audit in a refinery or petrochemical operation.

Finally, in Part D of Table 11.1, we encounter the ubiquitous area of miscellany. In Section 32, we evaluate noise and decibels, for which any safety professional has allowable limits and thresholds handy. Vibration and resonant oscillation are less direct (Section 33), and will usually signal a call for expert assistance through consultation with a structures analyst. Such a move by the safety auditor will normally be triggered by an obvious concern engendered by a visit to the operating area itself.

Section 34 has to do with fumes, dusts, and vapors—an Industrial Hygiene bailiwick. We recommend approaching this aspect by (1) determining what materials are actually present and/or used in the operation, (2) ascertaining what the MSDSs declare to be the operating parameters, and (3) obtaining from the Industrial Hygienist the data that supports or refutes the acceptability of the floor condition. If some or all of this data is unavailable, the safety auditor must immediately report the situation to operating management. For some multitalented professionals that are qualified in both industrial safety and hygiene, it may be like any other finding, and the finding, along with its corrective action, provided to management. Not all of us are so qualified, however, and part of our recommendation might very well be to involve an Industrial Hygienist immediately.

Section 35, on environmental extremes, will not apply to many industrial operations. Nevertheless, a dock worker in Duluth, or a refinery worker in Baton Rouge, might have occasion to question the propriety of conditions under which he has to labor. We present no prejudicial criteria for what "temperature extremes" are, but refer the auditor to the standards provided by the National Safety Council on the subject, and widely publicized throughout the nation for guidance in everything from ditch-digging to snowmobiling. The safety auditor should be aware that there are human limitations in this area, and when it is suspected that conditions are encountered that approach these extremes, the auditor should delve further, to determine whether a problem might exist. Another clue to this question might lie in the injury and illness records, if they are factual and detailed enough to provide such information.

Ergonomic considerations (Section 36) are a relatively inexact science. There are currently many seminars and short courses, and countless consultants offering their services, to define and correct ergonomic hazards. As an auditor, one is not expected to identify exactly which operational step, by which member of a crew, is a bona fide ergonomic hazard, unless the safety auditor is at the same time a qualified specialist in the field. The average safety auditor can only be on the watch for situations that appear to either strain the normal limits of people, or require repetitive exertions of an awkward nature. Whenever there is doubt about the long- or short-term effects of an observed or known activity, we recommend engaging an ergonomic expert to evaluate the situation. The questions posed in Table 11.1 are not intended to perform an ergonomic audit, but merely to jog the auditor's mind to be aware of the concern, and remind the auditor to seek expert advice when there are any questions.

Section 37 assesses the availability of emergency communication to others for the lone worker, or for the crew that is isolated together. It should not be forgotten that a group of a dozen can, collectively, be just as isolated and endangered as the operator of a one-man coal mine. They, as well as the single person, need a means of communication with emergency aid, with the rest of the world. The safety auditor must be aware of individual and group isolation and their need for "buddies."

Section 38, on confined spaces, is another part intended to jog the safety auditor's mind about the hazards of confined spaces, and not to exhaustively evaluate the confined space program in a given operation. The questions posed are intended to be answered on the floor and from the procedural review. In Chapter 10, the management evaluation checklist (Section IV, questions 10-11) ascertained the existence of policies and procedures for confined space safety. This section is intended to grade its implementation on the floor.

We have attempted to elaborate on the use of our operational checklist in plant operations audits. We started out applying the various points to our illustrative ink plant example, in order to show what the safety auditor might be attuned to in forays into the plant. We also discontinued this allusion midway through the checklist elaboration, because we feel we had made our point, and did not want to belabor the issue for those who are not conversant with, nor interested in, the details of a printing ink plant. Our objective has been to illustrate, in general terms, how the safety auditor, in any industry, can apply his or her experience and knowledge in that industry to the safety of operations there.

Generic Checklists

As we alluded to earlier, there are a variety of checklists that the Safety Auditor can use, depending on the objective of the current assignment. We have treated the comprehensive checklist thus far in this chapter, but we feel that a thorough treatment of the subject needs at least a cursory discussion of the application of generic checklists—those that are focused on a narrower field of hazards and problems.

In our ink plant example, reasonable generic checklists could obviously include such specific subjects as solvent storage, hazard communication, equipment and personnel grounding, and electrical code compliance. Equipment guarding and ergonomic adequacy are other subjects that come to mind in this hypothesized environment. We will take a closer look at the ink plant from the perspective of an audit of the conditions regarding a couple of these generic safety issues.

First, we should recall that an ink plant usually handles and processes many kinds of flammable liquids in equipment that is normally electrically powered. This includes blenders, pumps, mixers, and the ubiquitous ink mills. So, logical generic audits of the operation would be flammable solvent disciplines and the hazard communication program and practice. Let's look first at the former.

Table 11.2 is our version of a generic safety checklist for proper storage, use, and disposal of flammable solvents. These would include such widely used materials as acetone, naphtha, methyl ethyl ketone (MEK), and ethyl ether, commonly known simply as ether.

Table 11.2. Flammable Solvents.

I. Bulk and Drum Storage
 1. Are bulk storage tanks insulated and/or temperature-controlled?
 2. Are there adequate pressure relief devices for the tanks?
 3. Are the tanks grounded and provided with lightning protection?
 4. Is there an adequate spill collection basin?
 5. Is there an emergency action plan for spills or fires?
 6. Is there a fire-control system in place to prevent fire propagation from tank to tank?
 7. Are drums stored out of direct sunlight?
 8. Are they grounded and provided with lightning protection?
 9. If stored inside, is there adequate ventilation?
 10. Is there adequate spill containment?
 11. Are electrical devices and fittings appropriate for the Class I Division 1 environment?
 12. Is the area adequately covered for fire suppression?
 13. Are drums maintained in a grounded condition?
 14. Is there data to verify the continuity of grounds?
 15. Are tanks and containers properly labeled with hazards labels?

II. Storage Cabinets
 1. Are storage (use) cabinets of an approved design?
 2. Are the cabinets properly vented and grounded?
 3. Are the contents within the allowable limits?
 4. How many cabinets are in the area?
 5. Do containers in the cabinet have hazards labels affixed?
 6. Are they grounded where feasible?

III. Usage
 1. Are solvents dispensed for use in a safe manner?
 2. Can the person ensure him or herself to be grounded?
 3. Does the operator cross-bond from one container to another when transferring?
 4. Is the entire transfer grounded?
 5. Are Material Safety Data Sheets available to persons using and handling these materials?
 6. Is there adequate ventilation where dispensed and/or used?
 7. Is the correct personal protective equipment used?
 8. Is there data available for the area and work stations on the exposure, compared to the permissable exposure limit?
 9. Do workers return unused materials, or dispose of them, rather than leave them in the work area, when finished?
 10. Is the usage area adequately protected from fire spread?
 11. Is the area electrical equipment per the NEC?

IV. Disposal
 1. Are proper waste containers used for contaminated waste?
 2. Are waste solvent containers properly labeled as to contents?

Table 11.2. Flammable Solvents. (*Continued*)

3. Do disposal practices call for the separation of incompatible materials?
4. Are waste containers grounded, protected from heat and sunlight, and provided with fire and lightning protection?
5. Do disposal procedures comply with the MSDSs?
6. If inside storage, review Section I for drum storage.
7. At the end of a job, or at a suitable frequency, are waste flammables, and items contaminated with them, removed from the work area?

The ink plant will likely obtain these and numerous other flammable solvents in drums, cans, and bottles, but may purchase a few of them in tank truck quantities. It is likely that all of Section I of Table 11.2 will apply. In the ink manufacturing area, individuals will operate mixers and mills, often dispensing materials from the tanks or drums into smaller containers for use at or on the machinery. Flammable storage cabinets should be used for this working supply of solvents, and Section II should apply also.

As we have inferred, the work practices and facility characteristics evaluated in Section III come into play in virtually any usage scenario. Finally, disposal practices are equally ubiquitous, and will be evaluated from Section IV. We suggest that the flammable solvent checklist of Table 11.2 can be usefully employed in any operation that uses such materials.

Table 11.3 is a checklist for a hazards communication program evaluation, both from the standpoint of its physical embodiment, and its effectiveness on the operating floor. Again, we suggest that its application goes far beyond our ink plant. In fact, it was developed from Part 1200, Chapter Z of the OSHA Rules and Regulations, and as such, should prove applicable anywhere.

We recognize that Part 1200 is subject to revision and augmentation from time to time, and the checklist is not intended to be a substitute for knowledge and reference to it, at the time of an audit. However, the checklist should remind safety auditors to refresh their current knowledge before, or as, they embark upon a generic audit of a HazCom program.

Table 11.3. Hazard Communication Checklist.

I. Written Program
  1. Is there an existing written HazCom document for the company or department?
  2. Does it describe how materials are to be obtained, received, identified, and labeled?
  3. Does it require the supplier to provide Material Safety Data Sheets (MSDSs) for all materials?
  4. Does it detail supervisory duties and methods for informing employees of hazardous materials in their work area?
  5. Does it outline the necessary training elements of the HazCom program?
  6. Is there present in the work area a list of hazardous materials used in that area?
  7. Does the document show the supervisor's duties when unusual or infrequent tasks are to be performed?
  8. Does it specify procedures to be followed when nonemployees are brought into the area (contractors, visitors, vendors, customers, etc.)?
  9. Is the document available to all employees?
II. Labels and Warnings
  1. Are all containers labeled with the identity, hazard labels, and supplier/manufacturer's name and address?
  2. Are process vessels, containers, lines, vats, etc. that contain materials suitably identified?
  3. Are all such labels, signs, and markings legible and in English?
  4. Is there a need for additional labeling in a language other than English?
  5. Are there any OSHA-regulated substances present?
  6. Are they labeled in accordance with the specific requirements for that material, per OSHA Chapter Z?
III. MSDSs
  1. Is there an MSDS readily available for all hazardous materials in the area?
  2. Are they in English?
  3. Is the material identified as on the container label?
  4. Are ingredients identified where appropriate?
  5. Do the MSDSs provide physical and chemical properties, flammability, reactivity, etc.?
  6. Do they identify health hazards and exposure symptoms?
  7. Do they identify the route(s) of entry (oral, inhalation, skin absorption, etc.) that causes the health hazard?
  8. Do they give the PEL, TLV, or other exposure limits?
  9. If there are carcinogens, are they identified and noted as such?
  10. Do the MSDSs provide guidelines for safe handling of the materials, such as PPE and personal hygiene practices?
  11. Do they give emergency/first aid procedures?
  12. Are they dated and do they identify the originator by name, address, and telephone number?

Table 11.3. Hazard Communication Checklist. (*Continued*)

IV. Training
  1. Is there a training plan for the work area?
  2. Does it summarize the requirements of a HazCom program for the employees, in accordance with OSHA Chapter Z?
  3. Does it inform employees of the material hazards in their particular area?
  4. Does the training plan inform employees of the location of the hazardous material list and the corresponding MSDSs?
  5. Does it train employees in how to detect the presence of hazardous levels of materials, and what to do about it?
  6. Does it inform them of the symptoms and hazards of exposures?
  7. Are they trained in specific protective measures to be taken with these materials?
  8. Are they trained in reading and understanding hazard labels?
  9. Is there a record of which employees have been trained, and the content or scope of such training?

V. Products
  1. Are product MSDSs provided for each product?
  2. Are they complete, per Section III of this table?
  3. Are the containers of products ready for sale or shipment labeled, per Section II of this table?

VI. HazCom Program Effectiveness

   NOTE: We recommend interviewing a sample of workers and supervisors, to assess the general depth to which the HazCom program has been understood and implemented.

  1. Do foremen and supervisors understand the purpose and methods of the HazCom program?
  2. Do they keep their lists and MSDSs up to date?
  3. Do they have a training plan for new employees, for employee refresher training, and for nonemployees?
  4. Do employees know where the hazardous materials list and MSDSs are?
  5. Do they understand hazard warning labels?
  6. Do they use PPE properly and in accordance with the MSDSs?
  7. Do they understand the specific hazards associated with each material that they use or are exposed to?
  8. Do they know how to detect excessive exposures, and what to do when detected?
  9. Do they know the symptoms to be expected from exposures to the materials in their areas?
  10. Do they know the route(s) of entry that can prompt those symptoms?
  11. Do they seem at ease with their state of knowledge of the hazardous materials in their work area?

Field Notes

One can conjure up a humorous mental picture of the well-prepared safety auditor sallying forth into an operating plant armed with half a dozen checklists, OSHA and NFPA standards, management policies and operating procedures, equipment and layout drawings, and a note pad for writing down everything seen. Obviously this is not our recommended approach. It would, indeed, be almost impossible for one person, without a wheelbarrow, to haul all that stuff.

As we have discussed in Chapter 8, there are many volumes of reference materials that apply to a given operation, and they are, in fact, an essential part of the safety auditors' tool box. However, by the time auditors are in the plant, assessing the physical operation, they should have a pretty thorough understanding of the operation, its hazards, and what "should be." They may have derived a specific checklist for the operation.

Our recommended approach is to spend time in the area with only a note pad. Don't even take a checklist for the first several visits on the floor. Simply observe "what is," what is happening, what the equipment and people are doing, what the place looks like. Note clear instances of excellence, as well as obvious problems. Describe general overall impressions and specific items. As unobtrusively as possible, talk to people (that have the time), to learn what their job entails, how they do it, and how they feel about it. *DO NOT* take notes while you are interviewing operating people. It tends to make them respond artificially, if not defensively.

Back in the office, mental notes from the interviews and impromptu discussions, and field notes of observations and conditions, should be written up in an organized fashion, something like a diary. However, auditors should respect an individual's right of privacy when something negative has been disclosed or when an opinion about the work, or workplace, has been voiced that is less than glowing. The whole point of a safety audit is to discover facts that can be used to improve the safeness of an operation; it is not a vehicle for "getting" someone, or finding culprits.

After a few visits on the floor, and when the operations feel familiar, the checklist should be invoked. Remembering that it is a memory aid, and not the audit itself, safety auditors can use it to detail their findings, positive and negative, in a logical and organized way. If the comprehensive checklist is to be used, many sections will probably not apply. A simple "N/A" suffices to indicate that that section was not ignored or forgotten.

The visits without the checklist can be useful for finalizing a tailored checklist too. If auditors have opted for a tailored checklist, and based it on their homework from the drawings and procedures, their initial visits will likely result in some "tweaking" of that checklist, simply because of their coming face-to-face with reality.

With checklist in hand, whatever the type, safety auditors are ready to do the paper thing. We strongly suggest that each item on the checklist be completed *after* field notes have been assembled and assimilated. In other words, don't fill in the checklist — don't answer the questions on the floor. Instead, take notes, using the checklist as it is intended — as a memory jogger. If auditors are at the section on storage/racks/shelves (Table 11.1, Section 20), it is very likely that a lot of storage areas, parts racks, or tool shelves will be encountered in the operating area. We suggest that auditors make notes on each such installation, and complete their checklist evaluation in their office, where they can put them all together, and make an overall assessment.

The field notes are not to be considered a part of the record, in our opinion. They are to be taken as conditions warrant recording something, then compiled into an organized statement of the conditions found. The completed checklist is indeed a part of the record, but the field notes need not be. The reason we say this is that often, field notes made on the spot will reflect an element of ignorance on the part of the auditors. Subsequent education and illumination will often erase the question that caused the auditors to make the notation in the first place. In fact, we suggest that, once the field notes have been reduced to a coherent summary, and the checklist completed, that they be destroyed, simply because they have outlived their usefulness.

There is another dimension to field notes that we need to discuss. It is not unusual to canvass an operation, and find excellence in the handling and management of a safety matter, with one small exception. Say, for example, that an operation has twelve flammable storage cabinets, and the audit showed that ten of them were totally in compliance. The other two, however, were found to be overloaded, ungrounded, or of the wrong construction. The safety auditor cannot reasonably find the department unsatisfactory, but neither can the violations be ignored. The auditor can find the department satisfactory, and pinpoint the problem areas in the report. In this situation, we recommend retaining a list of specific findings that will become a part of the final report of the audit. This list will also become part of the audit record, but (as stated before) should be separated from the field notes. Once legitimate findings have been transposed from field notes to either a checklist or a final report list of problems, we still recommend getting rid of the field notes. They have outlived their usefulness, once they have been incorporated into a more formal vehicle for dissemination to management, which can cope with the issues raised.

Citing Requirements

The requirements for safety compliance have been treated at some length in Chapter 8. These are the yardsticks against which a given operation is measured. The source of the requirement can be extremely important to an operating manager. To him, a requirement that is imposed by federal or state mandate is far more compulsive than an advisory from NFPA or ANSI. Nevertheless, safety auditors must, in all conscience, cite the reference against which they make audit observations.

The pecking order has been alluded to before; federal, state, (and then it gets fuzzy) advisory, and company or industry rules and regulations. Some are incorporated by others, and some are advised by others. Last on the list of authorities for change from "what is" to what "should be" is professional opinion. However, we propose that, regardless of the status of the "requirement," safety auditors must identify its source. The impact that their recommendations will have will be directly proportional to the status in the hierarchy of requirements perceived by the operating management.

With this in mind, safety auditors should concentrate on those issues that are exactly equatable to OSHA and contractual (if applicable) concerns. This is not to suggest that advisory standards should be ignored, or that company or industry rules should be forgotten, or that professional judgment should be discarded. It merely proposes that, for the audit report to have the maximum impact and corrective action, the report must dwell most heavily upon the problems and issues that are most incontrovertible. If the credibility of the audit, and the auditors, can be established on solid grounds, such as OSHA, then the professional opinions offered will be much more readily accepted.

Regardless of the source, safety auditors owe their "clients" the reasons for their negative findings. If something is deficient, the "perpetrator" is entitled to an authoritative explanation as to why it is deficient. Knowledge of the applicable rules and regulations is obviously a distinct advantage in enhancing the legitimacy of negative findings.

Not to forget the importance of positive findings, we suggest that they be made both in the field and in the report. When there is notable compliance with a recognized requirement, and it warrants notation in the audit record, the authority establishing the requirement is noteworthy too. This two-way treatment of the acknowledged requirements will only enhance the credibility and perceived value of the safety audit, and the auditors.

Recommendations

The end product of a physical operating plant (or department) safety audit will be a written report. As we will discuss in Chapter 13 in more detail, the report should contain at least three elements. First should be the areas of compliance and excellence. Second would be the discrepancies, noncompliances, and problem areas that need to be corrected. Finally, and appropriate to both of the above, there must be recommendations.

The time to formulate recommendations is during the audit. If foremen show conscientiousness in the enforcement of the operating rules and procedures, it is worthy of notation as a positive finding. A corresponding recommendation would probably be that their supervisory methods be documented and emulated by other foremen on other crews and in other areas of the plant. The manner in which they achieve their excellent results may not be obvious to the auditors at the time, but the observable result—the finding demands a recommendation.

A clear deviation from an authoritative requirement demands an equally clear recommendation for compliance. It may also demand technical and engineering solutions to achieve compliance. It is not necessarily the safety auditors' duty to define the engineering or administrative solution, but it is their duty to define the discrepancy and its seriousness, and recommend that a solution be found.

It is also likely that auditors with some years of experience in the business, and some level of technical acumen, will, in fact, feel qualified to make a specific recommendation to correct a problem. They should not hesitate to do so, but they should not present their recommendations as the only acceptable ones. Instead, they should be offered as possible solutions, acknowledging that others might be equally or more effective.

Another good way to deduce and promote a valid recommendation is to solicit solutions from the crews themselves. This necessitates getting the operating people to understand what the safety problem is, why it is a problem (chapter and verse), and what the possible consequences of its continuation might be. With that understanding, more often than not they will have an excellent proposal for its correction. In fact, they may already be aware of the nature of the problem. If the auditors have gained their trust and confidence, they will not have to beg.

The recommendations that safety auditors propose must have a number of requisite qualities. First and foremost, they must correct the problem, or they must work toward expanding the level of excellence observed. Second, and almost as important, they should alleviate a *real* problem, and not be based on pure opinion. If they are based on professional opinion, then the tenor of the finding and recommendation should be so labeled. Finally, the recommendations must carry weight — they have to truly be technically sound in their purported role of correcting a real safety problem. The corollary of this is that safety auditors must depict the problem in an unequivocal manner, so that when they recommend studies by other technically qualified experts, they have nevertheless dispelled any questions as to whether the issue is real.

The recommendations emanating from a safety audit are its endproduct. The sponsor of the audit, whether management or client, has paid the auditor for nothing more than their expertise, their background and credentials, and their ability to find latent safety problems, and to suggest solutions to eliminate those problems. The energy with which safety auditors discover those problems must be matched by the energy with which they seek to solve them. The solutions may not all be theirs, but their genesis must be.

Chapter 12
Evaluating Work Practices

Observed work practices can be a very ambiguous area of auditing. What is seen on the floor, and what is written into the official procedures (if there are any), can be quite divergent. Yet, the divergence can have one of several meanings. We will undertake, in this chapter, to outline how work practice observations can be made, put into perspective, and dealt with.

Compliance with written procedures is wonderful, when encountered, but not necessarily the best for the safety of the operation. Auditors have to be conversant enough with the technology (or one of their pro tem assistants must be), to judge the adequacy and propriety of those written procedures.

We will discuss the procedures, how relevant compliance is, and how to make notes without intimidating the doers. In addition, the knotty problem of effecting management acceptance of the procedural deficiencies or deviations will be entertained.

This last is crucial, if improvements and corrections are to be engendered by the audit. The central message is to not indict the operating people for observed departures, until the reasons for those departures are thoroughly understood and evaluated. There might be very good reasons for the apparent procedural violations. Yet, to be fair about it, we acknowledge that there might not be valid reasons for such departures, and that disciplinary measures may be in order.

Procedures

There is no substitute for operating procedures. In spite of the fact that some of them go unread or unenforced for years, they are the stuff of which operating practices are made. Often, procedures are written and ballyhooed as the stuff of which dreams are made, but just as quickly evaded and short-circuited because of their impracticality. However, the operating procedures are the political and practical basis upon which audits, both the OSHA and inhouse varieties, are based. The ones that are in place at the time of the audit are the ones against which the operation will be evaluated.

Conformance to the existing operating procedures, at the time of the audit, is the subject of this chapter. Whether there is reason to question the adequacy of the procedures is also addressed, because a crucial question in any operational audit is whether the procedures are appropriate and correct for the operation involved.

Most operating procedures detail the sequence and way in which a particular step, or operational series, is thought to be best performed. Whether they describe the best and safest approach to the operational objective is the purpose of this phase of the safety audit. Safety auditors can only reach valid conclusions on this score if they have studied the operations, the MCEs, and the potential consequences of any MCE in sufficient detail to assess the procedures as they contribute to, or diminish the probability of, the MCE.

Besides the MCEs, there are bound to be many other events or circumstances that are negative, either because they cost money and time, or because they restrict activity aimed toward maximizing productivity. Try to remember that with us, the goal of safety professionals is not "Safety First," but rather "Success First." Success simply cannot happen if there is an accident that interferes with low costs, delivery timetables, or the acceptability of the product. They are so closely intertwined that they cannot be separated. The ubiquitous slogan "Safety First" should be a signal to any safety professional that the enterprise involved has a very shallow, and lip-service concept, of what true operational safety is all about.

Procedures should be evaluated on their merits. The correctness of a given operating procedure should be judged in light of the operational environment, the materials and products involved, and the nature of the equipment and facilities used to execute the operational objective. This is to say that, if the procedure describes the way in which the planner wishes to create item C out of items A and B, then the procedure for doing so should detail:

1. The hazardous properties and nature of materials "A," "B," and "C."
2. The precautions that one must employ when storing, handling, or using any of these materials.
3. The exact sequence of steps that are to be employed while creating "C" out of "A" and "B."
4. The hazards that one must be aware of when executing that sequence of steps or operations.
5. The methods of mitigating the hazards described in step 4 above.
6. The warning signals that will tell the operator that a potentially hazardous condition has arisen.
7. What to do when such warning signals are observed.

When the procedures are gauged against this standard, they can be adjudged as meeting minimum standards or not. Whether they are in compliance with *real* standards or not is another question.

Remember our discussion in Chapter 8? This is where we talked about the standards against which an audit should be made. Whether the procedures meet all of the criteria enumerated above is of moot value, if they are not in compliance with the appropriate standards for the operation in question. They must obviously meet both. If they fail our topical enumeration above, they are inadequate. Then they are clearly inadequate against published standards also.

The importance of written, authorized, and Safety-approved procedures cannot be overstated. They not only legitimize what is done on the floor, but provide safety auditors with the warm feeling that what is going on, on the floor, has been reviewed and approved by a fellow professional. If these procedures show no evidence of having been reviewed or endorsed by safety professionals, then auditors should consider them as virgin territory for their own professional evaluations. Even if such evidence exists, they should make their own evaluations, and deduce those areas in which they feel hazards have gone unrecognized, unabated, or insufficiently dealt with. When such questions arise, they should try to take up the issue with the colleagues involved. Often, those colleagues are no longer communicable (by virtue of geographical removal, death, transferral to a competitor, etc.) so the issue of second-guessing will rear its ugly head.

The safety auditors' role is, perhaps unfortunately, one of judging both the adequacy of the operating management's activity in regards to safety, and also that of the safety professionals who may have cognizance over those operations. It is a two-edged sword. Auditors may wish that no colleagues have been a party to the observed inadequacy of the procedures (or physical conditions, for that matter) that are observed. This will tend to relieve the safety community of any culpability for problems reported. At the same time, they may very well have to identify deviations, violations, or ignorances of existing requirements, approved by the safety professionals assigned to the areas. This will undoubtedly lead to tension with acknowledged colleagues, and the resultant cry of dissension in the safety community.

Safety auditors are nevertheless dedicated to the mission of reporting "what is," as compared to "what should be," and the findings that they have must be presented from that perspective. If the procedures are clearly in juxtaposition with what regulations and good practice demand, then they must, in all good conscience, report those discrepancies.

Procedures have several fundamental purposes where safety is concerned. First, they are to warn and instruct the "doers," of the hazards associated with the operations that they are performing. Second, they are to convey the correct way of performing the operation, so that the hazards are avoided to the maximum extent possible. Finally, procedures should instruct on what to do if the worst happens, and the operation goes awry—that is, a serious mishap occurs. Some may argue that this latter is the appropriate subject of an Emergency Plan, and that is an acceptable alternative, *if* the Emergency Plan is the subject of extensive training, rehearsed in detail and updated frequently.

It should be apparent to safety auditors, after a few floor audits, whether the procedures contain adequate instructions and warnings as to the hazards associated with the operations. Their floor observations, as compared to the identified hazards, from their "homework" (see Part II) and the written procedures, will force a conclusion on this aspect.

Auditors may or may not be competent to adjudge the "correct way" of performing the operation, in order to mitigate or eliminate hazards. Their job, then, is to communicate their concerns, in writing, to those who can, in fact, make those assessments. This is discussed in more detail in Part IV, to follow. Nevertheless, to the extent that safety auditors have or feel competence to question the technical and safety adequacy of the operating procedures in a given operating area, they must ethically do so. They can couch those challenges in whatever sweet words they choose, but the purpose is to reevaluate their safety-adequacy.

Whether the antidote to a mishap is in the operating procedures, or in some other document, whatever its label, is obviously not our concern in this discussion. What is important is whether the people likely to be present at the time of a mishap will know what response to make in a given circumstance. We offer, only as one of several options, that the emergency response procedures be included in the operating procedures, in which the operating personnel are (hopefully) well trained and rehearsed.

Operating procedures are, by policy and to audit standards, the way things are done. If they do not accurately define what is done; if they do not adequately describe the hazards and how to deal with them; if they do not unambiguously tell how, and with what, to accomplish the task; if they fail to instruct on how to cope with the unexpected (an accident); and if they do not tell the reader, or the doer, what the consequences of not following the procedures can be, then they must be judged deficient on one or more of those bases.

Compliance

In this chapter, we want to deal with the realities of compliance and what that means. There is real and visual compliance, and there is compliance contrived to pass a test, and there is compliance designed and intended to mitigate hazards. None of these is absolutely the correct one. The definition of compliance depends upon who is demanding it. Our theoretical treatment of the subject in previous chapters has been truly idealistic. We now acknowledge that there are numerous variations on the theme. That one manager's compliance might be another's overkill, and a third's absurd ignoring of the problem is the conundrum that we are discussing. The measure of compliance is a true function of what it means to the people who are contemplating it.

We hear this word (compliance) wherever we wander as safety professionals. To us, compliance is a very positive quality that we attach to what we observe in our plant visits. If we're not safety professionals, we may hear it in a different context, more of a threat then a goal. To safety auditors, compliance is a utopian state of affairs in which they can find no deviation from the expected norm. That is to say, there are no negative findings vis-á-vis the standards that they have found applicable to the operations that they are auditing. This seems to imply that there is no room for improvement, no way in which the operation can be made safer.

Of course it doesn't really mean this, because compliance is only the first step toward excellence. Reporting total compliance, when observed in a safety audit, can lead to a sense of complacency on the part of the operating management, depending upon the type of manager that we are dealing with. Safety auditors can then turn the lack of negative findings into truly positive ones, by dwelling on the positive fact of compliance, and perhaps challenging managers to proceed on toward excellence. Positive findings, as we have stressed before, can be powerful motivators for increased excellence. Whether an excellent observation is reported as such, with no negative findings, is a decision that safety auditors must make after they have assessed what type of manager is being dealt with (Chapter 2). Reporting total compliance is warranted when the managers are proactive and needing reinforcement in their safety endeavors. It is also warranted when the operating managers are the quintessential defensive type, or the type who push their own cause, but with a slightly different twist.

To auditees, lack of compliance is a very nasty word, and to be avoided at all costs. Their goals have to be to create at least the illusion, for the safety auditors, that they are in total compliance. Anything less is a potential blemish on their records in the company, and, as up-and-coming managers in the enterprise, a real impediment to their career ladder escalation.

What exactly is compliance? To the unthinking auditor's instinctively narrow point of view, it is the deviation from "what should be" to "what is." There are no gray areas. If the rope is supposed to be coiled clockwise, and the auditor finds it coiled counterclockwise, it is contrary to "what should be," and therefore a violation of whatever standard said that it should be clockwise. This auditor is probably doing no service to anyone, save to their own narrow view.

This type of auditing is less productive and less profitable to the enterprise than auditing that considers the hazards and consequences before pronouncing sentences of violation and guilt on the operating management, or the operating troops.

Our vision of compliance is much broader, and it presumes that safety auditors have become educated in the nature and hazards associated with the operations that they are auditing, as advised in Part II of this work (Chapters 5 through 8), entitled "Planning and Preparation." If auditors have not planned and prepared to the point where they are capable of distinguishing paper deviations from hazardous deviations, then their contributions in the audit will be of questionable real value. They may be of invaluable political value, however, as noted below.

Nevertheless, under these circumstances, such "paper deviations" can conceivably be worthwhile to management, especially if they are about to be scrutinized by another, outside agency, such as the scrutiny that the aerospace industry is subjected to regularly. Even federal or state OSHA scrutinization may warrent the shallow, pure compliance, type of audit that "unthinking" safety auditors might conduct and, eventually, report.

All of this is merely meant to acknowledge that the real and ideal worlds are truly two different spheres. What we have alluded to above is an acknowledgement of the incontrovertibility of the real world, while wishing for the ideal one. That is to say, there are times when idealism is intelligently set on the back burner in favor of stuntsmanship, in the name of "compliance." This is particularly true if the evaluator is less knowledgeable than the auditor in the hazards of the operation.

This has been a rather involved and likely controversial subject. What we are suggesting is that compliance is not a clear-cut black and white determination. There are "compliances" that meet the objective of eliminating or controlling hazards, and there are other "compliances" that satisfy external auditors, or evaluators, of such compliance. Among the latter, we must include the OSHA level inspections, where compliance may be measured in inches or parts per billion, rather than in hazard containment. Safety auditors employed by, or retained by, the management of an operation, must understand beforehand what kind of compliance they are expected to verify, or audit against.

A good example of this compliance distinction is the work practice that requires operating personnel to use supplied-air respirators while handling or using certain low-level permissible exposure level (PEL) materials. On the flip side of this "compliant" solution is the realization that if such PPE is employed, the workers are inhibited severely in any attempt to escape the greater hazard posed by (our imposed condition) an explosive material with which they are working, or using the low-level PEL material. The compliance issue that safety professionals place paramount might very well be the need to provide immediate and unimpeded escape from the explosive hazard. In that case, the use of a supplied-air respirator may be intelligently waived in favor of a tested and proven cartridge respirator that does not impede escape from the greater hazard, the explosive. While the supplied-air respirator meets the compliance dictates of OSHA, it ties the wearers, by virtual umbilical cords, to the operating area, and restricts their ability to escape — also an OSHA violation.

Compliance is not always a cut-and-dried proposition. Many circumstances will arise in which compliance with one requirement will be incompatible with another requirement. For safety auditors, though, there are two hierarchies of compliance. First of all, in auditing for the compliance of an operating department or area, there is the one level of compliance with the operating directives and procedures that are provided for the operating personnel. If they are operating in compliance with the directions and procedures that they are given, then their compliance is unassailable. However, if the directives and procedures themselves are out of compliance with higher-level requirements, such as the standards that we enumerated in Chapter 8, then those that produced the directives and procedures are out of compliance. Auditors have to be very careful about who is called to task about work practices that are not in compliance with standards. They cannot indict the doers, if the planners are in error. They cannot indict the planners if the doers are not complying with compliant planning.

Performance is either in accordance with operating standards, or it's not. The overall management of the operation may bear the ultimate responsibility for the noncompliance, and it may be a convenient ploy for safety auditors to assign the corrective action to requirements' noncompliances to that person, on the theory that the top dog is supposed to know all of the issues, and provide the coordination necessary to iron out discrepancies between "what is" and "what should be." We are not about to suggest a resolution to this question, because the politics of the situation are usually much more decisive than any theory or ideal that can be offered.

Sure, a lack of compliance is a management problem. Which management, is a local issue, depending on organizational structure and politics.

The deduction of whether an operation is, or is not, in compliance with the applicable standards and requirements, is not a simple "yes" or "no" question. From a purely legalistic point of view, it can be a fairly simple assessment. But from a hazard control point of view, it can be quite an ambiguous proposition. The nature of the organization and its political bent will probably dictate which type of compliance is expected. The fact that some type of compliance is required in the enterprise is a definite plus. However, legalistic compliance, compliance for its own sake, can leave nongoverned safety problems unabated with little hope of correction, simply because there "is no law against it."

Conversely, a compliance policy that demands control of hazards, whether regulated or not—simply because they are there, even for totally humanitarian reasons, takes the unfortunate chance that noncompliances that are also nonhazards will still elicit penalties that have no basis in fact, other than that they are noncompliant. But this is life! Safety professionals, whether auditors or not, must make their observations and recommendations in light of the world that they live in. It is not likely to be the one that they learned about in school, nor the one that we envision in this book.

Field Notes

When it comes to taking field notes for the purpose of observing and documenting work practices, we recommend that safety auditors *not* be too hasty in their zeal to catch workers "doing it wrong." Work practice observations are an inherently belligerent activity, from the standpoint of the workers. It has to be established at the outset that the auditors are in fact, *not* trying to catch them making mistakes, nor are they trying to find fault with the way workers execute their jobs.

In a facility and equipment audit, or in that portion of an audit that evaluates such, auditors can make field notes all they want, and do so in a way that does not intimidate the auditees. After all, the auditors are inspecting and critiquing inanimate things, such as floors, electrical installations, and cranes. Even in this phase of an audit, they should take pains not to relate deficiencies to individuals. Auditors are not there to find personal fault, but simply to find and report the faults and deficiencies, and the positive dimensions, of those inanimate objects. That this may reflect on a foreman or manager is a result that the auditors can neither control, nor alleviate.

However, when auditing work practices, the auditors are inescapably perceived as judging the adequacy, integrity, or purpose of individuals, or all of the above. The ones being observed cannot help but be worried about what the auditors might be finding wrong with what they are doing, or how they are doing it. This can then become, in the workers' minds, a personal affront, or threat, or attack on them. The situation calls for a different approach from that of the simplistic "what is" versus "what should be" that we have espoused in the previous chapters.

The first step in a work practices audit should be the establishment of friendly relationships with the operating people that are going to be watched and evaluated. This may sound trite and altruistic, but we can't help that. It is axiomatic that people level only with those they trust. That they must trust that the auditors are only interested in identifying hazards and helping to correct them, is essential to the success of this phase of an audit.

So our first suggestion is that, after having stated that purpose, safety auditors back it up with a thorough and sincere effort to know and understand what is supposed to be done to achieve a successful operation, *and* to know and understand the root reasons why the workers must do it otherwise in order to succeed. An all too frequent finding of a true work practices audit is that "what is" probably is superior to "what should be," as defined by the written procedures that are provided.

Auditors should strive very diligently to understand the common objectives of the operation, those that both workers and managers would agree to. In military terms, they should understand the mission. Then, they must have done their homework, and understand the hazards, as we have iterated many times before. Finally, the auditors need to know what can cause those hazards to create accidents, injuries and damages. When they have achieved that state of knowledge, they are in a position to intelligently assess the safety implications of both the physical conditions that they observe (the inanimate things), and also the actions and reactions that the workers live with to accomplish the mission.

When beginning an audit of work practices, safety auditors should explain the purpose of their presence in the operating area, and during normal operations. This introduction should be made in such a way that the operating people understand that they (the auditors) are seeking ways in which the operation can be made measurably safer. They should reiterate the known operating and material hazards, pointing out how and why they are hazards. This should be a factual kind of recitation of the obvious, and not a fear-provoking harangue on the unsafeness of the situation. If the MCE is a ruptured tank of a toxic substance, the auditors should simply remind the workers that a clear and present hazard is a tank rupture that would release a toxic substance. The very next breath should be used to also remind the workers that the tank and its appurtenances are designed to avoid such a rupture, and that the fail-safe devices engineered into the installation are also provided to do just that (if such is the case).

Having established that the tank of material is an inherent hazard, the auditors should go on to explain that their purpose is to discover whether there are other ways in which a toxic exposure might occur, other than equipment failure or inadequacy. It should be clear to anyone who is familiar with the operation that that kind of possibility exists, and that it is a good idea to know beforehand what activities and practices could cause such an exposure. The operating people may already know some of the ways in which they might encounter such a situation, and how to avoid them. If the element of trust has been successfully engendered, they will disclose them. The auditors need to build on this beginning by assessing whether there are other exposure scenarios, whether the practices that are in place to avoid them are correct, and more importantly, effective and followed.

To do this, it is note-taking time. The auditors can see without omniscience such blatantly unsafe practices as smoking while cleaning a part with acetone. They don't need a technical understanding of the operation to detect the unsafeness of working under a suspended load, or the storage of miscellaneous materials in front of an emergency disconnect switch. However, having noted such work practice problems, their note-taking should go on to determine whether this is recognized as a problem by those who are directly involved. A neat way of broaching this possibly ticklish issue is to say something like, "it seems like a problem to me that you are smoking so close to a highly flammable solvent. Does this seem like a problem to you?" The answer can be very enlightening. The person may never have thought of the possibility of being burned, or starting a fire; the worker may never have been told of the hazard, or instructed not to do so; or may not even believe in it.

Whatever the reason, the auditors should not attack the individual at this point, but (if necessary) leave the issue in the worker's mind that they (the auditors) are concerned. Education can come later, and most effectively, from the worker's supervisor. Of course, if there is an immediate and serious hazard, with the potential for catastrophic consequences, the auditors should take immediate steps to remove the cause and notify the area supervision of what they did and why. In the case of the smoker, they might simply move the solvent can to a safe location, if the supervisor or foreman is not accompanying them on the audit.

Field notes are the documentation of what the auditors observe. They should not be opinionated, nor accusative. They should simply state that at (time) on (day) such-and-such action was observed in (location), was hazardous (because), and so-and-so was informed, and/or (action) was taken. Later evaluation and research can expand on the degree of hazard, the reasons that it was accepted, and why. The note for the case of the smoker might simply state that an operator was observed cleaning with acetone while smoking, date, time and place, and that the solvent was removed, or the smoker asked to snuff it. The note might continue to the effect that the smoker's reply to whether he thought it was a hazardous practice, was "yes," "no," or "don't know."

There is no need, in most cases, to identify the offender. More than likely, further investigation will reveal that the worker was (1) not aware of the hazard, (2) had not been trained or instructed in it, (3) had forgotten about it, (4) doesn't believe there is a hazard, (5) thinks that a hazard that he or she believes they understand has been adequately circumvented, or (6) etc. All of these potential, and tentative, root causes for the hazardous work practice point to an underlying management cause, a failure to communicate, train, explain, remind and reinforce, convince, or discipline, and maybe more. The finding is that of an unsafe work practice, and management needs to take effective corrective action. Any added enlightenment that the auditors can provide merely helps them to pinpoint what type of corrective action that should be. This should be done without incriminating individuals.

Not every unsafe work practice will be as obvious as the previous rather simplistic examples. In many cases, the operating people will be conducting their processes and operations *exactly* as detailed in the operating instructions. Yet, the safety auditors, as professionals who understand the hazards, and what might cause one of them to precipitate an accident, are likely to equate an observed work practice to one of those hazards. These are the more difficult observations and findings to document and report. The safety auditors should probably not use themselves as authorities, against which to judge the adequacy of the written procedures. Rather, they should simply state the observed work practice and explain what they fear might result from it, and how it can happen that way, then refer the question (as a question, not a proven hypothesis) to an appropriate expert.

Other work practices may not be the possible direct cause of a hazardous situation, but might be in direct violation of one of the standards applicable to the operation, such as OSHA or the company safety rules. We pointed out previously that not all regulations are totally rational, but may have to be implemented anyway, depending upon the environment in which the enterprise operates. We hasten to add that regulations do not necessarily cover all rational hazards, either. In some cases, the goal attainment sequence may be compliance—then excellence—while in others, it may be just the reverse. But to the safety auditors noncompliant work practices are findings that must be noted too. The notes that they take are the documentation to management that there is, in fact, a compliance problem, and it is then their decision as to what action should be taken.

Regardless of whether the notation concerns deviations from patently obvious safe work practice, deviations from written procedures, faithful execution of a suspected unsafe procedure, or violation of an applicable standard, auditors should handle the note-taking with a measure of the following elements:

1. Discuss the concern with the people involved, explaining why it is a concern.
2. Ask whether it seems like a problem or hazard to them.
3. Try to understand their reasons for doing it that way.
4. Explain whether it is violation, a deviation, a suspected hazard, or a clearly real one.
5. Assure that you are going to get answers from those more qualified to evaluate.
6. Assure confidentiality, if it seems desired, but provide it, nevertheless.
7. Let the person(s) see the notes, and ask for confirmation that they are, in fact, accurate.

This last point goes a long way toward credibility enhancement for safety auditors. If the notation of the concern accurately, and objectively, reflects the actual situation that led to the finding, and does not incriminate the persons involved, they are very likely to buy into the discussion of the problem and its solution.

There are a host of much more authoritative works in the fields of human relations and communications that the reader might wish to avail himself of, in order to master the art of communicating negative information in a positive, reinforcing, and constructive way. We do not list ourselves among those authorities, nor do we offer a sure-fire formula for succeeding in this ticklish endeavor. We do know, though, that negative information that is conveyed in a factual and nonaccusative fashion will yield much more positive results than if it is not. Open objective observations, and equally open and objective notations, validated by the doers, will be more energetically corrected.

So far, we have dealt almost exclusively with negative findings and worries. As we have espoused so passionately before, the good things in life, the positive findings, are much more motivating, and likely to be received positively, than the negative ones. The note-taking must include documentation of every compliance observed, every cooperative action that the operating people gave, every observation of conscientious attempts to do the job correctly.

The positive findings are the stuff of which compliance, cooperation, and "doing it right" are both made and reinforced. The rapport that the safety auditors established early on must be preserved and embellished with findings of excellent performance. Auditors should never adopt the mind-set that some managers have, that doing it right is simply part of their job. This is such a demoralizing and punitive philosophy, that it will never foster willing and conscientious compliance. The positive reinforcement that comes from rewarding a job well done, can yield only one result — the desire and determination to do it right forever.

We have talked at great length in the preceding chapters about the use of checklists for management evaluations, and for assessing the adequacy of the physical plant itself. The use of checklists to evaluate work practices, and for augmenting or assisting in taking field notes, is a more difficult item to deal with. Work practices are "good" or "bad" from a frame of reference that is almost exclusively related to the kind of operation under audit. It is virtually impossible to provide a universal checklist that can be used to assess the correctness of operational practices in any and all industrial situations.

This leaves safety auditors with the perhaps odious task of creating their own checklists, or determining that they don't need them. We make this seemingly cop-out statement simply because safe and acceptable work practices are, in our view, virtually job-specific. What is important to the safety of a machinist's job will probably bear no resemblance to the safe work practices that will apply to a sewing machine operator in a clothing factory. We have compiled a partial list of job disciplines that we feel probably have some degree of universal application in general industry, as shown in Table 12.1. However, this is only offered as a nucleus from which safe work

Table 12.1 Work Practices Checklist.

1. How do operating personnel conform to written procedures for confined space entry?
2. Do operating practices conform to operating procedures for the handling and use of flammable solvents and materials?
3. Do people observe the rules on de-energizing and safeing electrical equipment before repairing or maintaining it?
4. Are areas around electrical or emergency switches and disconnects, and eyewashes and other emergency apparatus, kept free of clutter and obstructions?
5. Does the area warrant a generally clutter-free designation?
6. Do the workers follow the instructions for personnel protective equipment usage?
7. Is there a general observation of the zero-energy rule before servicing or maintaining equipment?
8. Do the operating personnel follow the procedures to the letter with regard to the operational sequence?
9. Are material-handling operations conducted with care and attention to the nature (fragility) of the material?
10. Are waste materials disposed according to MSDSs or other disposal instructions?
11. Are maintenance practices regular, and in accordance with established norms?
12. Do the operating people appear to be alert and aware of the operating environment in which they function?
13. Do operating practices, as reflected in the procedures used, acknowledge the hazards, and instruct on how to mitigate them?
14. Do personnel display an awareness of those hazards?
15. Do they follow the procedures, and report deficiencies in them?
17. Are machines kept guarded, even when frequent break-downs occur?
18. Are operating pressures, temperatures, durations, and limits observed as dictated by procedures?
19. Are Personal Protective Equipment (PPE) requirements, as delineated in procedures, followed?
20. Are the maintenance requirements for PPE enforced and observed?
21. And so on.

practices for a given operation should be derived. The specific practices will have to be derived from the hazards and potential causes that have been (hopefully) determined previously.

The items listed in Table 12.1 are not billed as inclusive, nor are they considered applicable to all operations. We include this tabulation as typical of the kinds of work practices that might be evaluated in an industrial setting. It is clear to us that operational work practice standards must be developed for the operation being audited, and we hope that the chapters on knowing the hazards and understanding them will serve auditors well.

Management Acceptance

Little can be added to our previous discussions about the role of management in the safety posture of an operating organization. We are firm believers in the premise that operating management is the central point about which safety, as a way of life, revolves. No one else can make it happen. Also, no one else can consign safe production to the trash heap as conclusively, and with such finality, as the operating management in a typical operational environment. It is this entity that calls the shots, and determines where safety really is in the hierarchy of operational priorities.

Whether operational management accepts, with aplomb or chagrin, the findings of safety auditors, those findings should be, as we have stressed before, assumed to be unassailable, and true, factual and objective. They should be presumed to be above question, from the factual standpoint. If auditors have found unsafe work practices, we must assume that what is reported is in fact what happened. The crucial question before us is, what does operating management do with this information?

There are a number of reactions that one might predict, depending upon the management style that the auditors are dealing with. Tyrants will rebel at the insinuation that one (or more) of their people are guilty of an operational infraction. They will move heaven and earth to prove that this observation is not true. Commandos will jump right into the fray, and start "corrective action" before they really know what the problem is, and so on!

Obviously, none of these "hip-shot" reactions are of any value in addressing the root causes of safety problems. A management that instinctively defends itself against neutral, objective observations is a management that cannot bridge the chasm between "what is" and "what should be." They are seemingly only interested, at most, in compliance, and not in excellence. If compliance is the only goal of the operating management, then safety auditors that identify noncompliances are the natural "enemy," and their findings are less likely to be accepted by those who are charged with managing the operation.

As we have stressed many times before, safety auditors cannot come on as critics that are going to correct all of the deficiencies that exist. They have to approach the audit as people who are willing to find and help correct safety problems. Management acceptance of their findings might very well depend upon how convincingly they have billed themselves, at the opening interview, as professionals who want to improve things, not incriminate people. There is no substitute for plausibility and trust.

We touched on this matter of trust in the last section, talking about the need to establish a trusting relationship with the folks on the floor, in order to fully understand what the real world is like. The same truth holds here, in that an element of trust is essential, between the safety auditors and the operating management, before they (the operating management) will accept the findings that the auditors offer. These findings will only be acted upon, if and when the management buys into their validity. Their buy-in will only happen when the safety auditors have passed all of the tests that, overtly or tacitly, the operating management have posted for them.

Management acceptance of audit findings ultimately depends on the cultural state of the operating management and, to a lesser degree, upon their management. It is no secret, and it has been a cornerstone of our treatise, that the safety conscientiousness of an organization is set by the top people. The CEO is a crucial player in this equation. His or her implied and actual stance on safe production will filter down to the operating floor, and management acceptance to audit findings will ultimately reflect the CEO's views. That is why we consider it so important for the audit reports to be at least copied to the CEO or equivalent in the organization.

The "head honcho" that is not attuned to the need for safe production, who doesn't understand the concept in the first place, will probably not be very supportive when it comes time to fund corrective actions. On the other hand, when auditors truly have the ear of the head of an organization, the corrective measures needed to alleviate safety problems will probably be supported and funded.

It is not our intent here to suggest that safety auditors should "suck up" to the top management, and forget the reactions that the local, operating management might have. This is a distinctly distasteful approach to publicizing findings, in our view. The first-line management must always be given the opportunity to respond positively to the findings of an audit, and to effect corrective action at that level. But safety auditors should, at the same time, remember that the local department manager is not the last word when it comes to correcting operational errors and work practice deficiencies.

Improvements

This section looks into ways to turn findings into improved ways of doing a defined job or process. We will discuss the more obvious activities for procedure compliance, such as training, educational approaches, process and equipment improvements, and discipline. We will stress the importance of making sure that management expectations are known and reinforced, as well as rewarded. We will also introduce the perhaps radical notion that maybe, just maybe, the operating folks know what is best, and the ivory tower engineers and planners might not be attuned to the real world.

The obvious objective of work practice auditing is to define ways in which unsafe work practices can be made safe, and then help to make it happen. When we were talking about physical plant deficiencies and noncompliances in Chapter 11, we arrived at a point in the audit where these were defined, and corrective action generally relegated, to management to provide the resources needed to design out the noncompliances. However, here in Chapter 12 we are concentrating on work practices, and their improvement or correction, so we must be about to assign to the workers the task of improving their unsafe practices—right? Of course, the answer is somewhere between an unqualified maybe and an unambiguous "it all depends."

Improvements in unacceptable work practices will rarely result from a legislative or authoritarian corrective action. Simply dictating "you will not do that anymore" will normally produce no lasting, and usually no meaningful, change for the better in behaviors. *That* is what we are really talking about when we consider improving work practices. Practices equal behaviors. Behaviors are the result of the total of all of the influencing factors under which people work—or try to.

When managers are informed of the unsafe nature of some observed operational practice, they need to know the fundamental "whys" before they can plan improvements. They need to know whether there is a training deficiency, an equipment or tooling problem being crutched, a procedural flaw such that the procedure cannot be followed as written, or a discipline problem needing their attention. Safety auditors may or may not be able to deduce definitive answers on their own, but the facts and observations that they provide to supplement the basic finding can be of immeasurable value to the manager, or their supporting engineers and planners, as well as the operating people themselves, in zeroing in on the root causes of the unsafe behavior or work practice.

Improvement efforts can be most successfully effected when the finding is presented in an unemotional and factual manner. The reader may well recognize this philosophy from prior chapters, since it is not peculiar to reporting work practice deficiencies. Any finding that has a negative connotation should be presented in that way. The finding should never incriminate, or editorialize, but should always connect the work practice, as observed, with one or more recognized hazards. In other words, "this" was observed, and it needs improvement (correction), because "that" can happen, and here is "how" it can happen.

Having reported that, truly proactive managers will probably take over and figure out for themselves why "this" is happening, and satisfy themselves that "that" can happen, and, finally, understand "how" it can happen. Remember, however, that while many operating managers are proactive and self-motivated to excellence, many are not. Engendering defensiveness is a poor start toward effecting safety improvements. So, on the chance that the operating management is not that proactive ideal, auditors should delve a little deeper.

That brings us back to the point of trying to assess the root cause, the reason why, the observed practice seems preferable to the operators, rather than the safe, correct way. We have opined before that people live up to their own perceptions of their bosses expectations. If in their hearts they believe that the foreman wants fast low-cost production above all else, they will probably move heaven and earth to achieve fast low-cost production. No other urgings will have much influence on their behaviors, as long as they stick to that premise. On the other side of that coin, if the operators believe that the boss expects sloppy work, because they think that they are considered incompetent, inferior, or dumb, then sloppy inferior work is the likely result.

One of the major stumbling blocks that managers run into is their own failure to articulate, convincingly and specifically, what they expect, and what will be rewarded. There are two elements involved in improving behavior patterns. The first, as we have said, is defining expectations, or job requirements. The second, and just as important, is rewarding the fulfillment of those expectations. At first glance, "reward" may come across as a grade school ploy. However, we are obviously not talking about a gold star on the spot, or an extra buck in the paycheck, or a beer at the local tavern. We are talking about what appeals to the inner self, the ego, if you will. There is not a person alive who doesn't glow at least a little bit when complimented, thanked, or appreciated. The sure-fire way to improve, change, and correct unsafe work practices is to make the person(s) feel good inside about having done it right.

There are several steps that need to be followed in order to arrive at that point where the good feeling can be imparted. We have touched on the first one already—identifying the real reason for the unsafe work practice. In the section on field notes, we discussed ways to interview operating and supervisory people about a perceived hazardous practice. Expressing their concern, and their own reasons for suspecting a hazard, safety auditors can turn the floor over to interviewees by asking whether they might consider the practice a problem too. This is often an effective opening for the kind of dialog that auditors need in order to gain some insight into why that operating practice is followed. Is it a shortcut that allows the workers to meet the time constraints that they feel pressured by? Is it simply easier, and their supervision has never objected? Is it because the written procedure is technically wrong and cannot be followed? Or is it because their equipment and tools don't perform the way they are intended to, and their supervision has not seen fit to repair or modify them? Are they reticent to identify a problem to a supervisor that, in their perception, at least, has a tendency to shoot the messengers? Also, of course, do they see no safety-related reason for not doing it their way?

Whatever the auditors find, and whatever the workers' reasons, auditors must be very careful to preserve confidentiality and (usually) anonymity, while carefully documenting the practice, its hazards, the rationale connecting the two, and the root cause for the practice.

Auditors should report this as a cohesive package, with the obvious goal of showing management where to attack the root cause. Nothing is ignored quite as readily as a problem that has no solution. So auditors have to deduce the reasons, even if they are, at best, suspected ones, behind the unsafe work practice.

The second step is in showing the true relationship between a cited work practice and a recognizable, or acknowledged, hazard. This is, of course, where safety professionals can cite requirements, experience, or engineering expertise. If there is no clear cause-and-effect relationship, even in the auditors' own analysis, then the work practice in question might very well be acceptable — *if* — the procedures or regulations allowed it. In which case, the work practice improvement might turn out to be a modification of the procedure or regulation that prohibits it. It should not be too surprising to safety professionals to find that the operating people may have zeroed in on the best and safest way to do a job, while the engineers and planners dozed away in their offices, oblivious to the real job to be done.

The point here is that work practice improvements may take the form of process and procedure improvements. That possibility must also be entertained, even though its acknowledgement can still be an indictment of the operating people that are doing the job better, but not per procedure. This is obviously a ticklish situation, as auditors want to help institutionalize the better, safer methods, and preserve the confidentiality of the operators, yet document and report the departure from authorized procedures.

Some managements will gladly accept the improvement that the unauthorized methods provide, and there will be no retribution for departing from "what should be." This will probably be rare, and not to be counted on. Auditors can get the most mileage, and improvement in work practices, by (1) maintaining confidentiality, (2) describing the procedural way, (3) illustrating how it creates or exacerbates hazards, (4) describing a superior procedure or practice, based on the conclusions from their observations, and (5) recommending changing the prescribed work practice to reflect that improvement. They do not really need to say that the "improvement" was what was observed. The auditors' observations can be the unsafe nature of the written procedure — period.

So we have seen that work practice improvements can take several forms. They can be effected through the finding of root causes and correcting them; they can be the result of illustrating, and convincing the operating people of, the hazards associated with their chosen way of doing the job; and they (the improvements) may be simply adopting and legitimizing their (the operators') own intuitive or evolved superior practice.

Recommendations

As we discussed in previous chapters, the recommendations that ensue from a work practices audit must be well documented, yet protect the operating personnel from potential retribution. We will now discuss ways in which auditors can present their findings without putting either the manager, or the crews, at risk from upper management, because of petty protective stances.

The recommendations from a work practices audit are necessarily controversial topics for discussion. There are few reports that operating managers will receive that will be laden with such potential indictments as an audit report. They know it will be circulated to their own management, and that it will reflect most specifically upon them. However, that does not preclude the possibility that the findings, and recommendations for those findings, cannot be made in a constructive and positive way.

The findings themselves are, in a manner of speaking, an indictment of the management practices of the particular operating managers involved. But safety auditors do have the opportunity of presenting those findings in a way that does not necessarily incriminate the managers. The recommendations that fall out of the findings can be presented in such a way that the operating managers are seen to be victims of the circumstances that have overwhelmed them. Whether this is an appropriate tack to take is left to the auditors, since it depends upon their evaluations of whether the managers are truly interested in enhancing safety, or merely going through motions in the name of safety.

Given that the auditors deem the operating managers to be truly interested in improving the safety of their operations, and receptive to constructive recommendations, safety auditors have a fairly easy task in front of them. Their recommendations will be accepted and acted upon with objective intent to correct. However, managers who react defensively and cannot accept the findings, and recommendations, as factual, will present a serious dilemma to auditors, because they will now face the "them versus us" syndrome. It is, at this point, incumbent upon safety auditors to document the issues (per the previous section "Improvements"), and overcome the natural reticence that operating management will have towards the realization that real hazards exist on their turf.

The auditors' main task in formulating recommendations is to relate them to the mitigation of recognizable hazards. In the previous section, we stressed the importance of relating observed practices to a recognizable hazard, so that the question of whether there was indeed a hazard, would not be a relevant issue. Now the emphasis must be put upon the question—does the recommendation directly address the accepted issue? In other words, does the recommendation point exactly toward the elimination of the hazard identified so eloquently heretofore?

If a perceived hazard is of little consequence in the overall scheme of things, if the MCE is fairly innocuous, and the worst thing that can happen is a skinned shin, then elegant corrective actions will likely fall upon deaf, or at least muted, ears. If the MCE is, however, indeed catastrophic, then the circumstances that might lead to the MCE are, or should be, of major concern to the operating management. So safety auditors must be aware of, and on guard against, "crying wolf," over hazards that are relatively inconsequential, and making iron-clad recommendations to combat them.

Recommendations must be offered in the context of what the true hazards are. Little is to be gained in the real world by recommending expensive solutions to minimal safety problems. The test of safety auditors comes when they can distinguish the primary from the secondary concerns, and focus their energies on the hazards that really need to be abated—now.

This is not to suggest that relatively minor problems should be swept under the proverbial carpet. The point to be made is that the recommendations should follow in approximate order of seriousness. There are few things that can detract from safety auditors' credibility than taking a firm stance on a relatively unimportant issue. Auditors must prioritize the issues before they compose their recommendations. Those that are of immediate and serious concern should be at the top of the list, while the problems that are of minimal concern will float to the bottom of that priority list.

As with any other aspect of industrial endeavor, the crucial must be separated (and discriminated) from the abstract, or trivial (though legalistic), issues. There are not too many problems that can be so distinguished, but they appear frequently enough to be a warning to auditors, that they should concentrate on the important issues and safety problems, and save the mundane compliance issues for a later evaluation. There are numerous "safety requirements" that are indeed trivial, but nonetheless critical to meeting the goal of compliance.

Depending upon the environment that safety auditors are working in, the relative importance of real safety problems, total compliance, and "getting by" will become apparent, and the auditors will have to adjust their focus to that real-world situation. The auditors' integrity, and the recommendations that their professional training direct, should never be an issue. The only decision that safety auditors, as professionals, should wrestle with, is the priority that they attach to the recommendations that they offer. Even the most innocuous and seemingly unimportant findings, as deduced from the requirements, must be reported, and recommended corrective actions offered. The emphasis that auditors place upon the corrective action recommendations will depend upon the MCE that the auditors can ascribe to the findings.

Chapter 13

Communications

Countless volumes have been written, and innumerable seminars and think sessions held, on this subject. There are probably enough works in existence to fill almost endless bookshelves, even rooms, full of the most expert thought on how to communicate effectively. So, how do we presume to devote a chapter, in this fairly technical how-to treatise on safety auditing, to the obviously unmastered art of interpersonal communication? We say unmastered, because it is our personal observation that an overwhelming percentage of the safety problems that arise are ultimately due to poor, inadequate, or wrong communication.

Our purpose in this chapter is not to rehash the practices and nuances that ensure proper and accurate communication. We will defer effective communications technique instruction to those rooms full of references that already exist in the literature on effective communication.

Our goal in this chapter is to explore and expound upon the message and manner of communicating safety audit findings to those who need to know. There is a somewhat unique characteristic to safety audit findings, as compared to other communication purposes, in that the audit findings are traditionally negative. If we don't leave another thought in this chapter, we want to reinforce the message of most of the previous ones—that audit findings do not have to be wholly negative.

The manner in which the negative aspects of an audit's total findings are presented is crucial to the way in which those negative findings are received, and acted upon. The negative elements will, however (if our advice thus far has been accepted), be far outweighed by the positive ones. It is the positive findings that auditors can make that should be at center-stage. The most important contribution that safety auditors can make is to observe, report, and reinforce good management practices, plant conditions, and work practices on the floor. The negative findings will, of necessity, be intertwined with the positive ones, but should never be the dominant portion of the audit report.

Communicating these mixed findings in a manner that is totally objective and factual, whether positive or negative, and with the requirements reference to authenticate them, is the challenge that safety auditors face when conducting audits, and especially when composing final reports. The level of effective communication that they displayed during audits is one factor; the manner in which they communicate in the final reports is another. Both must be done with sensitivity to the recipients. We will try to illustrate ways in which these communicative chores can be executed with the least pain, and the maximum benefit for all concerned.

Real Time

Communicating in real time is an extremely important element of communicating audit findings. Safety auditors who are fortunate enough to have operating managers accompany them will have no problem, if they can only speak. Describing and explaining the findings, as they are observed, can be a "piece of cake" if the managers responsible for the operations have enough time and interest to participate in the audits. The very fact that they are present might signal their proactive responses to the audit.

For auditors that do not luck out this way, it is still very necessary that they communicate their findings as quickly as possible. It is also equally important that those communications reach the eyes and ears of the individuals that can act upon them. Telling a floor lead-person or foreman, who has no authority or inclination to rearrange things to enhance safety, will accomplish little or nothing.

Real-time communication means informing those who can effect corrective action, and applaud and reward excellence, of the audit findings as soon as they are made. We depend upon the management to fall into step and do something, but this is true whether information is provided in real time, or in the distant future. Delaying information can only detract from the potential relationship between finding and correction, in the eyes of the folks on the floor, who see the hazard every day.

Especially when a finding is disclosed by those people, the value of immediate corrective action is inestimable. However, even if the finding does not originate from the floor, if auditors have discussed their concerns with the operating people, and they have understood the problem, fast and effective fixes are impressive. Not that safety auditors can guarantee, or accomplish, fast fixes, but they should not be seen as a delaying factor in the corrective action cycle.

The best method of effecting real-time communication is to ensure that the operating managers, or designees that have open and continuing access to the managers, know every finding daily. The findings should not only be communicated, but the rationale behind them. Why is it a finding? Why are auditors concerned enough to make it a finding? What is it about the system, managerial style, plant installation, or procedures observed that causes either concern or applause from the auditors? How and why did it catch their attention? These items of information should be imparted to the responsible management in real time, so that they might appreciate at once their importance in the eyes of the safety auditors. That these findings are important to the auditors may or may not instill a sense of urgency in the operating management, but (to the extent that that reception is perceived) it becomes another audit finding in the management evaluation portion of a safety audit.

In order to fulfill the goal of real-time communication, safety auditors must, as we have already said, tell someone, in some way, in a convincing manner, of the problem, and the reasons why it is a problem. That someone must be in a position to accomplish the needed action to correct the problem, or in a position to greatly influence the accomplishment of the needed corrective action. Real-time communication with the wrong or ineffective people is as useless as no communication at all.

This section could just as well have been entitled "Effective Communication," because that is the essence of our message here. However, that title sounded too trite and stilted, not to mention unimaginative. So we decided to dwell on the need for immediacy. To be effective, however, communication must clearly be both immediate and with the right people. The "what is," the "what should be," and the potential consequences of the disparity are the basic elements of such real-time communications. The recipient of these communications should be the one(s) who can close the gap, or eliminate the disparity.

Explaining the Problem

There is little more frustrating to professionals, no matter what their specialty, than making observations and recommendations to someone who has no concept of what is being discussed. The most important thing that safety auditors can (and indeed, must) do, when presenting their findings to the operating management, is to ensure their understandability. If the safety problem is not thoroughly explained, and conveyed to operating managers with total comprehension, then the problem will not be understood, nor corrected.

Explaining the problem, or (in another way of stating it) making the safety issue unquestionable—well understood—is the way in which safety auditors must communicate the safety issue to those who can do something about it. Clearly, auditors are in no position to correct unsafe conditions or practices on their own. So, it is only through their audit findings, and the understanding and acceptance of these by operating management, that auditors can hope to effect changes for the better.

This section is probably superfluous, since we have mentioned numerous times the need, and even the necessity, for real-time communication between safety auditors and the operating managers that are under audit. But we cannot overemphasize the importance of meaningful communication between the two.

Safety problems can arise from any of several sources. We have dedicated chapters to the three most elementary—management practices (as defined by management involvement and commitment), physical plant conditions, and work practices. While acknowledging that there are many cross-over situations that impinge upon more than one of these neatly distinguished categories, the cited problem, nevertheless, boils down to (1) a deviation from an acknowledged standard, or (2) a deviation from an adopted standard. The latter is the one that is most likely to cause controversy in the final analysis.

Explaining a deviation from an existing and acknowledged standard is relatively easy. "What is" can be provided from the audit itself. "What should be" is easily extracted from the standard that is being cited. There is a fairly clear-cut distinction between the two.

Explaining the problem when there is an element of opinion or judgment involved, or when the standard is not mandatory, as is the case, obviously, with OSHA standards, is much more challenging. This is where the value of the auditors' expertise and reputation, and the acceptance and applicability of the adopted standards, come to the test. Auditors that have no experience or expertise in the operations in question are probably better off, politically, to avoid opinionated findings. Explaining them from a relatively novice point of view is probably not productive. Also, safety professionals who blithely cite advisory requirements as though they were incontrovertible will not be very convincing either. Our advice, in trying to communicate this kind of finding, is to acknowledge up-front that the standard cited is advisory, or that the issue raised is based on knowledge and experience, and then phrase the "finding" as a request to "evaluate," "consider," or "determine" whether the trade-off is or is not in favor of compliance with the admittedly advisory standard.

There is, of course, an intermediate position that may arise, in which the safety auditors are respected members of the group, and their opinions and/or judgments do carry weight, irrespective of the "what should be's." It is then incumbent on the auditors to make doubly sure that their findings are accurate, and that the attendant recommendations are technically, as well as legalistically, correct.

Explaining problems can be a challenge, and the audit is not complete until the message is both sent and understood. That a problem might go uncorrected, simply because it is not adequately communicated, or explained, is completely unacceptable to safety professionals, and especially to safety auditors. The auditors' number one priority, during and after an audit, must be to communicate to operating management what concerns they have uncovered, why they are concerns, and how they might suggest that these be corrected. They must also remember that the final decisions on what to do about a particular finding are neither theirs, nor the technical advisors' that support them. They are the managers' decisions to make, and all that the safety auditors can contribute is advice and counsel, not decisive determinations.

To adequately explain the problem to operating managers, safety auditors must first get their attention. Like the mule that needs a 2 × 4 across the cranium to garner that attention, recalcitrant operating managers need some buzz-words that bring them out of the day-to-day coping with production problems, and make the current safety problems current production problems as well. Of course, most operating managers are not of the mule variety, and the 2 × 4 approach is inappropriate. But there are some "mules" out there, and getting their attention may prove to be a challenge. Normally, operating managers will understand the problems presented, even though they may not jump, immediately, on the bandwagon of correction. They may need time to assess the relevancy of the safety issues to their production goals.

The cultural metamorphosis that accompanies the equation of safety problems with production problems is not an easy thing to effect. There are, in fact, numerous references that attempt to make that correlation. We do not propose to rehash those theoretical hypotheses here, but merely to reemphasize that the need to communicate effectively with operating management is paramount to effective auditing.

As auditors, safety professionals are nevertheless bound to the ethical standards espoused by the professional, legal, and company standards for which they have signed up. To do this, they have to explain the safety issues to the operating management, in a manner that is as unambiguous, incontestable, and incontrovertible as possible. This requires a total understanding of the hazards, the MCE, and the governing regulations, on the part of safety auditors. This, then, leads us back to Chapter 5, on knowing the hazards. If auditors do not truly understand the operation and its potential hazards, they cannot communicate their concerns effectively to the people that can correct them. They cannot be very convincing, identifying perceived hazards in operations that they barely understand.

The need for effective communication with, and its acceptance by, operating management is paramount. The most effective way of accomplishing this goal is to be totally familiar with, and understand the technicalities of, the operation, and then to communicate with that operating management throughout the audit, and never hesitate to explain a safety problem. Doing so in real time is far more effective than springing surprises on them when the audit report is issued.

This last point cannot be overemphasized. Many safety problems, whether compliance issues or work practice hazards, can be corrected virtually by edict, or with minimum pain in the budget. One of the most palatable findings, from the viewpoint of both auditors and auditees, is "corrected on the spot," or "corrected during the audit." It is in the best interests of both auditors and operating managers that findings, and suspected findings, be communicated to the managers as quickly as possible. Immediate corrective actions can be acknowledged, and neither party comes out tarnished by the label of being insensitive to safety problems. Of course, all bets are off if operating managers prove to be completely disinterested in the safety of their operations. The only alternative then left to safety auditors is to report the problem(s) and let the chips fall where they may.

Acknowledgements

Findings are made in three basic ways. There are those that safety auditors (1) deduce on their own, based upon requirements, legal or internal, or (2) professional evaluation, based upon their own understanding and knowledge of the operation and its hazards. The third source of safety problem findings, and often the most significant one, is the insight that comes from the operating personnel themselves. They are undoubtedly the most knowledgable people in the world, when it comes to knowing and understanding the nuances of both the technical and safety aspects of their operation.

Safety auditors must always remember that they are not the all-seeing gurus of the operations that they are auditing. They are, rather, outsiders who are offering their observations and advice to enhance the safety of the operations. It is an unequivocal observation of fact that auditors will "discover" less than the operating people already *know*. Auditors are not going to make determinations that are wholly revolutionary to the folks that run the operation. Any finding that safety auditors propound will be old news to the people that run the operation, unless they are totally oblivious to the hazards they live with every day. The auditors are not going to discover new facts regarding the hazards of the operation. They can, at best, deduce the problems and dangers that the existing hazards (known to the operating people) present. The learning experienced, in the course of the audit preparation and execution, can only serve to bring auditors closer to the level of understanding that the operating folks already have. However, that very understanding can, and should, provide auditors with the tools needed to illuminate the hazards in terms that the operating management will take seriously.

So, since safety auditors are not likely to rediscover the world, but are thrust into the role of discovering the problems that are present in a given operation, they must learn to present their findings in such a manner that they will be "news" to the operating management. The findings should be novel enough to engender a level of interest that elicits action, yet couched in terms that the operating personnel will recognize, and endorse as worthy of management action.

The issue of acknowledgement is a potentially touchy one. If the issue is too parochial, too closely identified with a particular operating department or individual, then its impact can be diluted and the identified parties subjected to unnecessary critique by the next level of management. The most important element of problem identification and acknowledgement is the maintenance of individual anonymity, while at the same time elucidating the programmatic implication of the problem. This is to say that the safety problem should be illuminated without making "bad guys" out of the people that live with it every day.

Indeed, the "bad guys" might be the sole source of the deduction that a safety problem exists. It is wholly irrational for safety auditors to indict their most informative colleagues, so the acknowledgements must be made in a manner that validates the authenticity of the finding, preserves the anonymity of the source, and provides reward to the revealer. This is not easy, and usually one or more of these attributes has to be sacrificed, in order to report the finding in the first place. We suggest that the primary consideration is normally the anonymity of the source. This is not to say that there may not be occasions where the authoritative value of the source, or the incentive of an acknowledged disclosure, may not outweigh the value of anonymity. This is a judgment that only safety auditors can make at the time they are writing up the findings.

There is an inherent value to acknowledging the sources of safety problem identification, but there is the flip-side to be reckoned with. Acknowledgement and identification are not necessarily compatible. Auditors should make sure that the sources of their understanding and eventual findings are comfortable with their role as "informers." Otherwise, the auditors might better forego the authenticity that the "informers" offer, and report the findings as their own.

The Final Report

This document is the product of safety auditors. There is nothing else that they will do that will have more, or less, impact on the safety of an operation than the drafting of the final report. It can be the impetus to management to effect the changes necessary to accomplish a truly safe operation. It can also be the vehicle by which Safety, regardless of its various hats, is relegated to relative insignificance.

The nature of audit reports is almost one of tattling. Of virtual necessity, the operating management's superiors are copied, if not addressed, in the report. In order to avoid the tattler's role, safety auditors must be extremely selective in their phraseology. They not only must maintain their total objectivity, but also avoid outright accusation and indictment of the operating folks that they have been working with for the past several days or weeks. This, of course, presumes a negative premise—that the audits have found fairly serious safety problems that could not, or were not, corrected quickly.

This, then, raises a very important point, which we alluded to in the last section. If a safety problem, however profound or trivial, is truly corrected during the audit, it is extremely important that this fact be acknowledged and applauded in the final report. As we said earlier, this is one of the most palatable tasks that safety auditors can face—the "conversion" of previously oblivious operating managers to those who truly attempt to improve the safety of their operations. Never mind, at this point, whether the corrections are conscientious or superficial. The strokes for having done something meaningful just might fire the zeal of managers who were on the safety "fence."

If the strokes of the audit reports are reinforced by the operating managers' superiors, the value is, of course, multiplied. There is hardly a person alive that does not bask, just a little bit, in the accolades of another, regardless of the latter's status. That very "basking" will have an effect on the managers' safety consciousness —take our word for it.

The final reports, which safety auditors must consider as public as the six o'clock news, even when the distribution is quite limited, should state each and every finding—both positive and negative— with the detachment of an IRS examiner. Auditors have, in theory, no allegiances toward anyone. They have only the compulsion to expedite excellence in the operations that they have cognizance over. Altruistic as this all sounds, there is a ring of truth to it, and the final report should reflect that ultimate objective. That is, there are no "bad guys," and there are no rotten situations. There are simply things that are. They may be physical conditions, procedures that are used or systems that exist, but they are "what is." They are not what some scoundrel created.

To illustrate our point with another in a continuing series of simplistic examples, suppose an auditor finds that there is no positive, physical lock-out procedure for the Maintenance Department. The auditor can report this finding in an infinite variety of ways, but let's compose two that are diametrically opposite in terms of their objectivity and detachment.

1. Management has failed to provide, communicate, or enforce a lock-out procedure for maintenance on energized equipment.
2. The auditor was unable to find a lock-out procedure for energized equipment maintenance.

Need we say more? The first is a direct indictment of a sleazy management that could not care less about the safety of their people. The second is a fairly bland statement of the finding—that a procedure could not be unearthed. Both demand the same corrective action—the production of a lock-out procedure. However, think of how much less animosity the second finding engenders.

That is what safety auditing is all about—the improvement of the safety of an operation, not the conviction of violators.

In previous chapters, we have talked about recalcitrant managers who refuse to accept or acknowledge the safety deficiencies over which they reign. We must now acknowledge that the situation will arise where auditors must, in all professional conscience, report that recalcitrance. Let us again phrase two such findings, where the lack of a lock-out procedure is the finding, and the responsible manager has expressed absolutely no inclination to abate the deficiency.

1. When confronted with the noncompliance that a lock-out procedure was not in place, in spite of OSHA dictates that one be in existence, and enforced, the maintenance manager scoffed at the need or practicality for one.
2. There was no visible enforcement of the mandated requirement for a lock-out procedure.

Again, the first of these is pure indictment, whether deserved or not. Even if true, no real productive purpose can be served by this kind of reporting. The second version conveys the same deficiency without incriminating, or questioning, the ulterior motives of the management. It simply states the floor observation that the lock-out procedure was not employed or required, and leaves it to the management team to determine how, and indeed, even if, compliance will be effected.

One of the truths that safety auditors must assimilate, and reflect in their reporting, is that they are not enforcers—they are only reporters. Regardless of the glowing credentials that they may have had bestowed upon them, or the lack thereof, they are not the last word in safety in that operation. The auditors are advisors, working at the pleasure of the management. If they are to have any lasting influence, and not engender hostility in the area, their reports must reflect the findings in a neutral, nonaccusative, and totally objective way.

Other Audits

We have concentrated quite exclusively on the industrial safety aspect of safety auditing. That was, and is, the sole purpose of this work. At the same time, we wish to acknowledge that there are other, equally important, elements of an audit program, and that those may very well be the ones that dictate the make-up of a safety auditing program. We are well aware of quality and process audits, and recognize that these disciplines contribute greatly to the overall excellence of an operation. Whether they can assess the safety issues, in as thorough a fashion as we have tried to depict here, is another question.

We know of quality and process auditors who are the essence of excellent and objective observation. It doesn't take a CSP to recognize safety problems. What it does take to identify safety problems is a working knowledge of the operation itself. This talent is clearly not the exclusive property of safety professionals or safety auditors.

The kernel of this section is very simple, that dedicated safety auditors should acknowledge and accept the fact that others, of varying disciplines, will be auditing also. We should acknowledge at this point that there are quite common driving forces that motivate auditors of any and all labels.

All auditors bear the totally negative label of *auditor*. Whether they can outlive that wholly defeatist beginning is a reflection on the humanistic qualities of rookie safety auditors themselves. To be able to overcome a negative, but virtually automatic, label is a true measure of the safety professionals' professionalism.

There are complete courses currently available on Industrial Hygiene audits, environmental audits, and many other specific aspects of the overall "Safety" discipline, or loss prevention, field. It would serve no positive purpose here to rehash the principles espoused by those other professionals in the field. Suffice it to say, while we are concentrating most heavily on industrial safety in this work, the other disciplines cannot be ignored. To auditors that are looking for either excellence or deficiencies, it is entirely possible that some of their findings will touch on those other disciplines. When that happens, we beg ignorance, and refer the reader, or auditor, to a higher authority in that field.

That safety auditing should be so parochial may be a surprise to some readers, and we apologize if the scope of this work is narrower than what might have been expected. We reiterate that our baseline of discussion was that of industrial safety professionals who are primarily interested in maximizing the safety of their cognizant operations. That we cannot include specific sections and chapters on environmental and industrial hygiene aspects is a fact that we must acknowledge as a limitation of this work, and by such acknowledgement, refer the reader concerned with these aspects to the experts mentioned before. We are attempting here to cover only one major element of a total audit program.

While the other elements might very well be as critical to the success of the operation, we can only refer the reader to the discussion in Chapter 1, where we strived to make the safety, quality, cost, and timeliness of production all part of a whole. Auditing the success of that perfect production objective must, of necessity, evaluate all of those facets of success. A truly synchronized audit function, which is aimed at deducing the degree of compliance to all the operating requirements, will include more than safety compliance. It will evaluate all of the relevant issues, including safety. This work is an attempt to provide guidance only in the area of Industrial Safety. Loss-prevention professionals will realize immediately that they have numerous other disciplines to look into, before they have come up with the comprehensive audit plan for all possible risks.

PART IV
Analysis

Findings are one thing; what to do with them is quite another. Safety auditors are inevitably going to have scores, hundreds, and eventually thousands of individual findings. After having reported them in their audit reports, and extracted some promise of correction in "budget year 19XX," or a procedural change by November 12, 19XX, the obvious question is, "so what?" What are safety auditors supposed to do after they have reported the safety problems, or excellences, resulting from area audits?

In this Part, we will turn to the issue of analyzing and, more importantly, reporting the results of those analyses, of the safety audit findings. Individual findings are, of course, important, to the degree that they do, indeed, identify safety issues. Some of these may, in fact, be critical to the safe production of the operation. Others will, undoubtedly, be of a seemingly minor nature, but raise the question of whether they are representative of a trend, a tendency toward "getting by," that is indicative of a potential threat to safe production. However, the real, effective use of audit findings is in the distillation of individual findings into generic issues that portray to management where their lapses are.

The core of this Part is in the deduction of an atmosphere. The purpose of any analysis is to deduce the general from the specific. The general conclusions from audit findings are the atmosphere that we want to deduce, and the individual findings are the various weather conditions observed at different points in time and place. The countless specific findings can lead only to an equal number of specific corrections. However, the reduction of those numerous specific findings to a manageable handful of generic issues can lead to the definition of the atmosphere, and the categorical solution to most of those countless safety problems, in terms of a relatively few, real, and meaningful corrective actions.

Safety auditors owe it to the operating management to reach as low a denominator—a bottom line—as they can. This Part will explore ways in which the audit findings can be reported, and analyzed, to make the operating manager's job as uncomplicated as possible. After all, the safety auditors' mission is not to enumerate some quota of findings; it is to reinforce what is best in the operation, to convince management of the need to correct what is worst, and to consider the trade-offs implicit in the compliance questions that are somewhere in-between.

On the other hand, the auditors' mission is not necessarily to make life easier for managers. Nor is it to make life more difficult. Their mission is to inform—to provide quality information that the managers can use in the execution of their job of producing products most effectively and efficiently. The totally objective analysis of audit findings is the cornerstone that can validate the findings, individually, and stimulate broader, more comprehensive implementation of correctives on a generic scale.

This last point is the essence of truly effective safety auditing. When operating management "sees the light," and actually recognizes that there are shortcomings in the overall way in which they do business, then "across-the-board" corrective actions are most likely to occur. If they are lulled into the false security of believing that problems are unique to Department "C," or "Plant F," then the corrective actions will only address the specific issues identified in Department "C" and "Plant F." This is not the goal of effective safety auditing. The ultimate objective of safety auditing is to change, or reinforce, the culture, into one that seeks to understand the risks and exposures, and control them. At the levels we are dealing with, this is best accomplished through the generalization of the specific.

So let's go on to see how we analyse the numerous findings, both positive and negative, that repeated audits are bound to reveal. There is no substitute for reality. And, there is no substitute for real-world examples. The ensuing Chapters of Part IV will use a lot of real-world examples to illustrate the analytical approach to making sense out of countless audit findings. Remember, when an operating manager buys into a generic safety problem, it is likely that dozens, if not hundreds, of individual safety problems will be corrected.

Chapter 14

Evaluating Management Performance

Long-term trends, the stuff of which performance analyses are made, is a completely ambiguous concept. Whether long-term is from one fiscal quarter to the next, or from one year to the next, or even decade to decade, is a matter for local definition. In some relatively ponderous industries, such as railroading or mining, long-term changes may be measured in years and decades. In other (more dynamic) industries, such as specialty chemicals or food products, long-term trends might be determined over six months or less. Changes and improvements in safety culture and accident rates are tied closely to the industry standard for evolutionary change. If an operating management opts to be a leader in its field, then change can, of course, occur more quickly than the industry norm might suggest.

The previously stated objective of safety auditing is to instigate or reinforce a management goal of perpetual improvement in safety performance, not through gimmicks and slogans, not with banners and memorandums, but through evolutionary change in the industrial culture. To maintain thrust in this direction, management needs continual information that focuses their attention on those areas and issues where performance improvement is most needed, whether because of present performance (the accident record) or catastrophic potential (the sleeping and uncontrolled Maximum Credible Event).

First, however, management needs to "buy in." They must acknowledge that a serious risk is out there, and, having been made aware of it, that they are themselves at risk if they ignore it. It is not news to safety professionals that many cases have been won by plaintiffs who showed present knowledge *and* a lack of management action to alleviate hazards. This is why it is so important, during the audit itself, that auditors cite their source and authority for each finding. As we will show a little later in this Part, these sources are just as important in stimulating generic solutions to generic safety problems. Dangerous inaction often comes from a lack of appreciation of the true exposure created by the problem — exposure both in terms of immediate loss when an accident occurs, and the potential litigation that might follow. We do not intend to dwell at any length on legal entanglements, but simply to recognize that they are also a serious risk in today's society. Many works exist that treat the liabilities of management and supervision, as well as the company, when acknowledged hazards go uncontrolled. The issue is relevant here only to the extent that safety auditors make it crystal clear to the operating management what the issue is, why it is an issue, and what the potential consequences are if it remains uncorrected or uncontrolled.

This point expands geometrically in importance when we are dealing with the generic issues deduced from repeated findings of the same, or similar, violations. Repeated and wide-spread violations of safety requirements are viewed as much more serious than the occasional violation missed in an otherwise proactive compliance program. When a pattern can be discerned, safety auditors must immediately communicate that pattern, and their individual findings that have led to its discernment, to the responsible manager.

Management activities and actions are the subject of this chapter. It is not going to be easy, or necessarily comfortable, for safety auditors to make generalized observations and conclusions concerning managers' safety stances and performance. There is bound to be the natural reticence to accuse, or indict, a fellow human being. We cannot overemphasize the idea that the individual findings, as well as the concise distillation of them, should never be accusative or indicting of an individual manager. They must always be fact-by-fact recitations of "what is," and compared to "what should be." When phrasing findings of a negative nature, safety auditors should conscientiously seek words and phrases that are painfully objective and impersonal, and stay away from statements that convey blame or dereliction of duty.

When the findings, and the analytical summaries of those findings, can be communicated in such a manner, the chaotic results of accusative reporting can be avoided. The issues themselves can be addressed without coping with the questions of who is implicated, who is guilty, and who is none of the above. Situations discovered and reported by safety auditors are *never* the result of a single individual's dereliction of duty. It is *always* the result of a management system's failure to encourage and reward appropriate behavior and procedures.

Twisting the Tiger's Tail

This will seem (to many readers) a curious, if not ridiculous, title for a section of a book on safety auditing. It originates in a long-ago experience, in which the author was an initiate in a well-known service organization, and the entire point of the initiation was to make the initiates most uncomfortable. The "tail-twister" was an elected official whose job was to find any fault he could with the novices, whether it related to the dress, conduct at the meeting, or utter lack of anything to criticize on the part of the initiate. The tail-twister was an autocratic entity, whose authority, and power to instill discomfort, went unchallenged.

He was regarded as the one person to keep on one's friendly list, even though he had no constitutional charter to treat his fellow-members as the dirt he professed to ascribe to our worth in his world. As young novitiates in our early twenties, it never occurred to us that perhaps his power over us was the result of our own definition of what he was, and what influence he really had.

Twisting a tiger's tail is a very risky activity. Not many of us are really willing to do this either, even if we have had the experience of dealing with the tail-twister. A real tiger will most probably react with some measure of displeasure, if his or her tail is twisted to the point of discomfort. If the tiger doesn't snap your head off, it will be equally apparent that he or she doesn't like the bother of such indignities.

The parallel we are trying to make here, is that the tiger becomes indignant, rather than hurt or defensive, when someone has the audacity to twist its tail. It suggests that the poor old tiger has no other option but to deal with the tail-twister. Whether the twisting has any merit has been lost in the tiger's reaction to the twist itself.

Safety auditors are, to one degree or another, tail-twisters, simply because they are cast in the role of critics. Calling the local operating management less than perfect is a risk similar to twisting a tiger's tail. Either activity is bound to result in "disagreement" from the "twistee" — the one whose behavior, or activity, or program (vis-a-vis safety) is called into question.

We want to offer here a few suggestions as to how safety auditors might deal with having the unsolicited role of outsider critic — the tail-twister. The first idea that we might offer is quite obvious — don't act out the role of a tail-twister! Don't be the one who comes on as the unassailable pain-causer. Safety auditors are not, and should never bill themselves as, the final authority, unchallengable founts of knowledge, that dictate what must be. If they assume the role of tail-twister, they will undoubtedly be confronted by many foes (tigers) more knowledgable than they, who will tear them to shreds (figuratively) for their audacity in criticizing the operations under audit.

Therefore, we recommend that safety auditors analyse their findings in a manner that avoids the role of hard critic, and rather presents them in the vein of incontrovertable facts. There is no preferable option to objective reporting, and analysis. Does this sound familiar?

This means that auditors must separate their facts from their prejudices. More than that, it means that they must validate their findings, before pronouncing a verdict. It also means that they must not be afraid of pronouncing conclusions on generic problems, when the individual findings lead inexorably to conclusions regarding those generic problems, and the same set of findings fail to show a generic treatment of the problems.

Generic solutions to safety problems are clearly one of our cornerstones, from a philosophical standpoint. Editorially speaking, we do not believe that symptomatic solutions are very worthwhile. It is only when individual symptoms (findings) are generalized that meaningful corrective actions are instituted. In order to realize comprehensive corrective actions, "tigers" must come to understand that (1) the problems exist and are real (the MCEs and their supporting data), (2) their exposure, both morally and legally, is real, and (3) there are solutions.

Once that realization takes place, that there are real risks with real solutions, there are few operating managers who will opt to junk the whole idea, and revert to business as usual. Nine times out of ten, or perhaps more, safety auditors will have made their point, when they have elucidated the risks and their relative priority, and solutions or recommendations. In this environment, there is no real need to twist the tiger's tail. The only essential element of such auditing is that the tiger knows as much as the safety auditor does.

Credibility

The essence of the safety auditors' mission is credibility. It matters not what they see, find, report, or campaign for, if they do not establish credibility, they will be about as worthless as a seven-dollar bill (inflation has taken its toll). The credibility of the safety audit findings is paramount to effective safety auditing, as we have stressed before. The way in which such credibility is established is to make sure that no finding has *only* an element of professional opinion, or *only* a sterile regulation interpretation, to back it up. A finding must be translatable into the real world of the operating floor, the plant itself, before it can take on meaning for the operating manager and crews involved.

The point that we are making here may seem obvious, but it is crucial. Safety discrepancies are only corrected when they have touched the heartstrings of the management concerned. The only way that safety audit findings (describing safety deficiencies) can be corrected is when the operating management concurs that the risk is unacceptable, and that *something* must be done. That something might well be what the safety auditors have recommended in their reports. However, we hasten to add that it need not be the safety auditors' solutions that save the day. Auditors must always remember that their solutions, their recommendations, might very well be the least optimum of a whole variety of corrective measures that can be taken to alleviate a safety problem. As we have stressed before, no one knows the problem, or its solution, as well as those who live with it daily.

The credibility—the reputation—that safety auditors bring to the analytical process is even more important than the credibility that may be attached to any given individual finding. As we have mentioned before, the individual finding might, in all probability, be corrected in totally acceptable fashion, and appear to have been resolved. However, the professionals' credibility is what stimulates operating management to aggressive implementation of safety problem corrections. Unfortunately, novice auditors, or experienced safety professionals in a new role, may not bring that credibility into the initial audits. Thus they must revert to the axiom that each finding must be thoroughly documented against a standard, and professional opinion may not hold much water for a while, until professional credibility has been established.

The transmission of safety problems to other areas of the department or plant is often rare, and is one of the reasons why an analysis of findings is beneficial. It is clearly valuable and important that findings be transmitted from one operating area to another. It is even more important that those findings be acted upon in a universal, conclusive manner.

Suppose (our first promised example) that, while auditing the paint storage area of a metal-finishing department, the auditor finds that there is no lightning protection system, and that there is no provision for grounding the metal paint cans. Is this a real safety problem? For the safety professional, the former might very well be considered a serious deficiency, and the latter a negligible one. Our technical assessment, incidently, is that it is far more effective to protect the entire storage area with a proven counterpoise system, than to try to ground a multiplicity of individual containers. Others may have different technical views on the matter, and some organizations may have deduced, and made into procedure, quite specific standards for this situation.

The point is not so much what the solution is, but the identification of the problem itself. If the totally ungrounded and unprotected condition found is either (1) contrary to the stated standards of the organization, or (2) contrary to safe practice, as promulgated in the NFPA codes, then a problem has been identified. If the auditor finds similar situations in the paint room of that same plant, or if the painted parts in the drying ovens are found to be just as isolated electrically, then the problem takes on a much larger magnitude, and suggests either a deficiency in internal standards, a lack of understanding of the application of any applicable standards, or perhaps a disregard for the relevancy of the standards. A more global problem becomes apparent as the auditor evaluates the growing data base of audit findings.

The solution may then shift from a local correction of a grounding problem in a single paint storage area, to a more universal policy statement regarding the storage and use of flammable materials throughout the plant, or the company.

One obvious way to get a head start on this issue is to make findings known throughout the organization, regardless of how parochial the initial finding might be. This is nice in theory, but frankly, we would not bet the mortgage that a plant manager in South Carolina will launch a corrective action campaign, based solely on the news that a problem of a similar nature was found at a sister plant in Kansas. We hasten to acknowledge that there will be those laudable exceptions where such action will be undertaken, but do not expect this to be the norm.

The next thought is to inform the common management of both, or all, "sister" plants, of the problem, and recommend an umbrella policy that will have force and effect throughout the organization. When employing this tack, safety auditor needs to be sure that the specific problem becomes translated into a generic one, and that the policy solution is generic as well. The auditor should not recommend that a lightning protection envelope of X feet by Y feet be provided for Building Z. Rather, he or she should recommend that a policy be established that the lightning protection envelope for any flammable storage area extend W feet beyond the dimensions of the structure. This generic recommendation then applies to the entire company, and does not leave the local manager, in whose operation the problem was first identified, as the guilty party, and lone actor, in effecting corrective action. As a bonus, if this approach proves to be effective, the problem does not have to be rediscovered numerous times, to become a universal one, before a policy correction is edicted. This approach allows the safety auditor to generalize from the specific, almost immediately.

The credibility factor assumes its importance in the original documentation of the problem. If we are going to successfully achieve this "instantaneous" generalization, then the presentation of the original finding must be unassailable. The regulatory and/or technical facts have to be accurate, and the situation (the local finding) meticulously detailed. This takes more time than a superficial documentation of a specific deficiency, and the auditor must make the trade-off early on, whether to report specific findings and make equally specific recommendations, or to translate local findings into policy matters, and thus approach the safety problems on the broadest possible front. The organizational politics will likely be decisive when making this determination.

Either way, however, whether trying to prompt correction of a problem in one square yard of one department, or institute across-the-board solutions within the total organization, accuracy and objectivity are still paramount. The credibility of the finding, as it compares "what is" with "what should be," cannot be sacrificed because the field of view is narrower. Successful generalization at the outset only serves to reduce the number of times that the same issue must be dealt with.

The decision we alluded to in the preceding paragraph will probably be based on the "politics" of the organization, which is an evasive way of saying that it will probably depend mostly upon whose ear the safety auditor has—who has commissioned the audits in the first place. If it is the CEO, the generic approach will likely be quite successful, providing (of course) that the auditor jealously protects accuracy and credibility. If his or her mentor is the safety manager at a local operation, then generalization is probably the wrong choice. The auditor will then have to build up a catalog of specific findings across the organization, showing a pattern, attempting to ascertain root causes and MCEs, before presentation of the data needed to attract the attention of the CEO or other executives that can reinforce the generalization. Unless the top management is fast on the uptake, and this does happen occasionally, building the case for policy level action will be more tedious. However, the local corrective actions, taken one by one, will be rewarding to the auditor as well. The auditor has to learn to deal with the world in which he or she functions, and not curse the world wished for—simply because it might not exist.

Have we departed from the title of this section and chapter, which we billed as stressing credibility when Evaluating Management Performance? We suggest not, because management performance has to be evaluated in that real world in which we all function. The safety auditor has to understand what compels the operating manager's actions. The auditor's performance is likely determined by the resultant of a number of disparate forces. Like any force diagram, the magnitude and direction of a particular force determines what influence it has on the reaction of the body in question—in this case the body (and mind) of the manager. The manager's response to safety audit findings is inevitably a result of the manager's desire for safety for his or her people, the maintenance of the ability to meet production goals (minimum property loss), profitability, productivity, self-preservation, and personal career enhancement (all disparate forces acting on the manager). If the manager is to react to safety problem correction with the fervor the auditor might wish, then the consequences of inaction should be couched in terms that speak to as many, if not all, of these driving forces, as can be brought to bear. The safety auditor can enhance the performance of the auditees by interpreting his or her findings in terms that stimulate and coordinate all of the forces that motivate management.

The essence of Chapter 17 is analysis itself. In the next several chapters, on evaluating management, physical, and work practice deficiencies, we intend to elucidate what the goal and limitations of such analyses can be in the real world. We ask the reader's indulgence, meanwhile, while we attempt to philosophize on the purposes of analysis, before examining the nuts and bolts of Chapter 17.

Analysis without purpose can be quite unproductive. In a sense, all deficiencies are a commentary on management practices, in that they reflect a condition that is tacitly, overtly, or helplessly accepted by the operating management. One of the very subtle challenges to safety auditors is to try to determine which of these might be operative. Not because they can, or should, report their conclusions on the question, but so that they can understand which level of management they are evaluating. If the acceptance of safety problems is done at the floor managers' level, then the conditions clearly reflect that level's safety performance. The managers either don't understand or believe the problem to be real, or don't care, or are oblivious to the goals, regarding safety, of upper management. However, if the floor managers' stance is that of a clear inability to influence correction of the problems, then safety auditors are evaluating higher levels of management. It may not be immediately important which case exists, but if a pattern emerges, the conclusion will be incontrovertible. When top management espouses one standard, and middle and lower management perform to some lesser standard, it is the latter's performance that is in question, and the top levels need to know where the deviations are occurring. If the top management is truly espousing safety, not just with words, but with deeds and expectations, then this information will be acted upon. Also, when middle and lower management are in conformance with top management's standards, and safety problems still exist, then it is the top management's performance that must be addressed.

It is a circuitous evaluation, attempting to pinpoint where (at which level) the conflicting forces are steering management into adopting or accepting unsafe operations. As we said earlier, it is not the safety auditors' purpose to report such deductions; it becomes useful information when composing their factual observations for audit reports.

We feel this to be an accurate assessment, whether the specific findings deal with policy, system, physical plant condition, or observed work practices.

Recommendations

The formulation of recommendations, from safety audit findings, is a seemingly simple, straightforward, exercise. The fact is that safety auditors, while presumed to be safety professionals capable of recognizing system, management, physical, and procedural problems, are not necessarily systems analysts, management consultants, facility or tool design engineers, or process engineers, as well. Their recommendations cannot be expected to always be the optimum. In some complex instances, they might, indeed, have no specific recommendation at all; they may simply recommend that a competent specialist in the applicable discipline study the problem and recommend an optimized solution.

This inevitable truth has been a problem for the safety community for as long as there has been a safety community. It has been common, over the years, for the Safety PERSON, whatever the title or station in the organization, to bless any and all designs and activities. It has been just as common for the Safety PERSON to be the goat when the design fails, or the procedure doesn't work as planned. The reason is obvious; the Safety PERSON was obliged by *management policy* to approve all such documents, whether he or she knew anything about tool design, stress analysis, or chemical reaction kinetics, or not.

What safety professionals will ever be technical experts in all of the technical disciplines involved in a twentieth-century production enterprise? Clearly, few to none! The most that safety professionals of any stripe can offer is the challenge to designers or process engineers to show them how, why, and under which conditions, proposed plans or designs are safe. If safety auditors are capable of validating the designs, failure points, or reaction equilibrium of a dynamic reaction, then they are in the wrong department.

For these several reasons, safety auditors should not, generally, offer technically restrictive solutions. If a technical recommendation is to be offered, illustrating a novel approach or idea, it should be conceptual only, and offered for evaluation by a competent specialist in the applicable discipline. Too often, and we have been guilty of this also, auditors are convinced that the problem, however incontrovertible, has only one solution — theirs. This is the surest way to lose credibility, and to have recommendations be totally ignored.

Of course, we are talking so far about technical — not regulatory, problems and solutions. For our second example, in this chapter, suppose that the safety auditor discovers that there was an air quality analysis run three years ago, that showed levels of toluene in a work area to be half the permissable exposure limit (PEL). Further, nothing has been done in the way of improved ventilation, vapor capture facilities, or discharge controls. Yet the consumption of toluene has, in those same three years, more than doubled in that operating area. The safety auditor is now hot on the trail of a negative finding, and indicating management for dereliction of duty in not having a system, or not using it, to assess this apparent problem.

The safety auditor can conclude one of a number of findings and make a corresponding number of recommendations. He or she can flail out and report that no follow-up air quality studies have occurred for three years, and therefore a serious health hazard exists that must be fixed with vastly increased ventilation. Alternatively, the auditor can observe that, while toluene consumption has increased dramatically, no current data on concentration is available, and a sampling program must be instigated.

In this simplistic example, it should be apparent that the latter is the professional's course of action. It eschews the emotional jump from partial fact to concrete correction. It does not leave the auditor open to the obvious challenge — where is your data that shows this upgrading of the ventilation system to be necessary? The clear choice of recommended action is, instead, to get the data that either confirms the need for improved ventilation, and defines how much improvement is needed, or determines that the exposure is still within current limits. Few can challenge the wisdom of obtaining current data; but many will challenge improvement for altruistic and unjustified reasons.

This example, quite simplistic of course, illustrates another vital concept. If the safety auditor recommended doubling the ventilation rate (with no exposure data to support the recommendation), and subsequent analytical data showed that the exposure was actually three or four times the assumed concentration, based on usage alone, then the safety auditor can be liable for a faulty recommendation, because it fell woefully short of adequate for control of the hazard. When the auditor dons the cloak of technical expert, he or she has to do so with the utmost caution, recognizing his or her own exposure to liability. By acting on his or her suspicions, and recommending a sampling analysis to define the exposure, the auditor has shifted the onus of incorrect action, or nonaction, onto the responsible management. This is clearly proper!

By definition, recommendations are made to those who are responsible, and can act upon them. They are not made to the world at large, nor to the operating crews, the budgeteers, the schedulers, or the local Chapter of ASSE. They must be aimed, in terms that can be understood and acted upon promptly, at those who would manage. The auditor has to give them an option that addresses the problem at hand, yet maintains their inalienable right to manage.

Couching recommendations in this vein will not always be simple. When repeated problems of a similar nature have been identified by ongoing analysis, it becomes increasingly more difficult to refrain from becoming very specific. If a pattern is established, safety auditors must, in all conscience, refer the generic issue to upper management, as we described in the last section. But whether specific or generic, recommendations should be made in terms that allow facts to determine corrective actions.

We are treating, specifically, the analysis of management performance, their systems, dedication, and activities regarding managing their safety responsibilities. Safety auditors cannot make a declaration of their performance until they have measured that performance against the standards that are to be met, and this is achieved from a valid sampling of that performance over a period of time. It can be a very difficult transition for managers at all levels to make, from thinking only of efficiencies, schedules, scrap rates, and manpower needs, to (in addition) assessing the adequacy of the equipment and planning to conduct the operations without loss. To those managers that show a positive effort and progress toward the integration of safety into production planning, their performance must be considered exemplary, and recommendations couched in terms that will encourage, not deflate, that effort.

For managers whose interests lie mostly in avoiding bureaucratic hassle, by minimal compliance only, the evaluations can only assess their performance as compliant, if not proactive. While for managers that flaunt the rules in order to achieve large production figures, while hazards go ignored, their performance is clearly inadequate. However, it should not be the safety auditors that publish these possible conclusions. It must be the managers' managers that conclude these performance evaluations, based upon the documented findings of the auditors.

Going back to Chapter 10 for a moment, we discussed a number of measures of management performance — involvement, commitment, activities, employee perceptions, systems and their utilization, and delegation. Whichever of these various element were reviewed, and findings documented, established a baseline (if a first audit) or depicted a movement, one way or another (if a follow-up). An early indication of progress, or its lack, can be seen in the reply that the operating managers make to the initial audit. If follow-up shows a concerted effort to formulate and implement activities and systems that lead out in the quest for safer production, then auditors can report these findings in as factual a tone as they did the original findings. A point-by-point comparison of each item in the checklist will serve to illustrate, in tabular form, where the improvement has been made, and where greater emphasis must be placed.

In our view, this kind of impassionate analysis of the managers' safety performance improvement is important in helping auditors avoid the role of critic. It allows progress, as well as regression, to be disclosed without becoming judgemental. At the same time, we endorse giving plaudits openly when conscientious effort and change are found. A compliment for genuine effort in improving, or integrating, safety is rarely unappreciated. Integration is, remember, the primary goal, not improvement in an isolated vacuum of "Safety First."

The bottom-line message of our discourse on recommendations regarding management performance is this: tell it like it is; contrast their documentable performance against the standards that they are bound to; disclose whether they acknowledge the existence of those standards; report those instances of performance above and beyond the requirements of the standards, however complete, correct, superficial, or minimum they might be; and applaud the activities that exceed those minima. Further, if progress can be discerned in the proactive direction, note that and applaud it, as well.

Chapter 15

Identifying Physical Deficiencies

This is the part of safety audits that many people will equate to "safety inspections." Safety auditors are out to nail the operating folks to the wall, and there is little or nothing that operating management can do to withstand the "corporate snoops" that are there to find out everything bad about the operation. Safety auditors, of course, must do what they can to dissolve this perception, and substitute the idea that the safety auditors' role is one of technical, regulatory, and compliance assessment. This is far more easily said than accomplished.

The physical deficiencies uncovered by in-depth safety audits of operating areas are only indicative of a management condition. We state this somewhat categorically, firmly believing that "what is" is the product of what management allows, expects, or demands. Nothing that exists on the operating floor is there because the operating management is, or has been, oblivious to "what is." Whether tacitly or overtly, the physical conditions that exist are the accepted standards by which the operating people function.

The purpose of this chapter is to elucidate the methods and protocols by which the identification of plant deficiencies can be most fruitfully reported to those who can correct them.

Documentation

When reporting physical deficiencies, safety auditors are both professionally and literally obliged to possess documentation of the discrepancies that they are about to publish. At the same time, in order to avoid being labelled as whitewashers, they must be in a position to document any positive findings reported, also. We have dwelt at some length on the former premise, but feel obliged to elucidate a bit on the latter. Positive findings are, by nature, suspect in some skeptical management circles, especially those that are of a dictatorial bent. When tyrant or commando managers see a report of positive qualities in their work forces, they are likely to rebel at, or at least question, the authenticity of the findings.

We want to reiterate the importance of reporting and, therefore, documenting both parts of the body of findings. There is little corporate benefit to be realized from a negative finding that has no concrete evidence to reinforce it, nor is there any profit to reporting a positive finding that cannot be validated, or at least supported, by outside evidence.

The documentation process begins with the floor audit or, in some cases, with the management systems audit. If the deficiencies are attributable to the systems in place (or not in place), such as hazards communication, confined space control, accident reporting and recording, etc., then the physical deficiencies revert to management deficiencies. The lack of a management system for dealing with such fundamental safety issues is clearly a problem with the management system. Any system deficiency is, in our view, a direct reflection of a management deficiency. Whether the specific item is a policy shortcoming, a physical plant deviation, or an operational discrepancy, it is likely to be traced to a management system deficiency.

However, we are dealing (in this chapter) with identifiable deviations from the "expected" management performance, the comparison of "what is" with what management expects, or demands, and "what should be," in terms of the physical plant standard, and the supposition that the safety auditors find something quite different. Now what?

The crucial question is, "does management *really* expect something different from what they have? The answer to this vital question is an extremely critical one, and nearly impossible to provide with any degree of certainty. The conundrum for safety auditors is, does the operating management espouse one set of standards (what should be), and encourage, condone, or even create another (what is), or does management truly rue the existence of what is, and feel helpless to correct it? Even more to the point, is operating management aware of the safety problems that exist in their areas?

If management is truly convinced that "safe production" is the way to maximize success, and lacks the ammunition to make it a way of life, then the safety auditors' findings might very well be the spark that ignites a desire, throughout the operating area, to improve and correct the deficiencies noted. If, on the other hand, the operating management is indeed two-faced about safety, and only mouths the tenets during "Safety Awareness Week," then the particular findings of a safety audit will be ignored, or at best, rationalized away. It is the safety auditors' burden to distinguish one from the other, and determine which stance the operating manager is coming from.

Having been there, the authors recommend steadfastness in safety auditors' position. That is to say, safety auditors should never report anything that is not factually reinforced, either by "requirement," or by technical evaluation and validation. There is no substitute, in management circles, for the informed opinion, statements of requirements, or both, in influencing management toward the safe decision. Few indeed, in today's environment, are those managers who would flout the advice of their professional advisors, or knowingly fly in the face of indisputable and authoritative requirements.

Yet the safety auditors' advice and recommendations, pertaining to a particular safety problem, will be challenged, scoffed at, even acted adversely upon. If the safety auditors have, indeed, documented the finding, and its counterpart in the requirements, they need not fear the reporting of same. At the managerial levels that most safety auditors will operate, the resistance to physical recommendations will undoubtedly be fierce. How often have we heard retorts like, "we have been doing it that way for fifteen years, and you're the first one who ever wrote it up," or "that machine has always been unguarded because we have to make a lot of adjustments. We can't be always taking it off and putting it on again."

Documenting physical deficiencies means more than just taking notes, however detailed, while in the operating area. It demands an understanding of the technology well enough to distinguish the differing levels of hazards and priorities, knowing the business environment—the forces that drive both management and operators, and the resources that can be brought to bear on solutions. Documenting and reporting minutia to the virtual exclusion of the higher-level hazards can be either misleading, or tend to erode credibility, or both. When reporting physical deficiencies, safety auditors are both professionally and literally obliged to possess documentation of the discrepancies that they are about to publish (sound familiar?). The bottom line here is that safety auditors must know what is required and convincingly report what should be done about it.

The basic premises involved in documenting physical deficiencies are (1) ensuring that the problems are real, that is, survivable against an assessment of "what should be" versus "what is," and (2) that the observations are totally credible. It does little good to document a safety problem in a situation that is atypical, if that situation is not representative of the operating mode under which the department functions. The key to genuine physical conditions' assessment lies in the discovery of what obtains during "normal" operating conditions, and contrasting those findings with the standards under which the department is supposed to operate. On the other side of the coin, we caution that the prepared situation may not be atypical, at all, of what is normal. The point for auditors is that they evaluate the operation in as nascent a condition as reasonably possible.

To this end, we suggest that safety auditors "invade" the operating area under audit, to establish (as best they can) what the allowable limits are. That is to say, if the acknowledged parameters are at one level, and the tolerated performance is at another, then the physical deficiencies may indeed revert to another management deficiency. Finding what really exists when no one is looking calls for a bit of subterfuge.

However, if the physical deficiencies noted in the safety audit are indeed products of the professional expertise of the safety auditors themselves, then it is incumbent on the auditors to label their findings as such, and, more than ever, document the contrast between "what should be" and "what is."

Documenting physical deficiencies is not a readily accessible game. There are many things that will tend to interfere with this seemingly simple dictum. Among those most obvious are the political realities of the actual situation that safety auditors find themselves in. If they are in their positions to truly inform management of the safety problems extant in the area, then the reporting of departures from those requirements will probably cause no particular chagrin on the part of the operating management, or the auditors' themselves. However, if the environment is such that the bearer of bad news (problems) is in fact the problem, then the safety auditors are in line for some difficult times.

This potential adds greatly to the case for incontrovertible assurance that the findings of safety auditors are totally factual, and will "hold up in court." That is to say, these findings should be of such a nature that disputation is not a viable option for someone who wishes to challenge the validity of a particular thing.

In order to "affect" the maximum "effect" on a safety audit program, or a safety auditing element of such a safety program, safety auditors must document every finding, both positive and negative, that they discover. Documentation involves a number of elements, which include (1) the observation of what is compliant or noncompliant, (2) the statute or advisory standard that is or is not complied with, and (3) what the safety auditors suggest to rectify the two, presuming noncompliance.

We will deal in more specific terms with the translation of deficiencies into action in Chapter 17. In the meantime, it should be quite apparent to the reader that the identification and categorization of physical plant observations are central to a true safety audit of an operating area. Whether the findings are substantive or trivial, the safety audit will portray to the top management what the safety priorities are in this particular operating area. Whether this information is of interest, and acted upon, is a wholly different issue. There is no substitute for management involvement, the desire to make the operating area as safe as can be done.

However safety auditors have no control over this element. They can report what they find, and they can recommend what to do about it, but they *cannot* effect change or improvement on their own. They must rely totally upon their factual reporting.

Endemic or Epidemic?

One of the most striking things for an outsider, who must nevertheless spend a sustained period of time in an operation, is the isolated departure from what that person has learned to expect in that plant or area. These unexpected islands may be of either the positive or negative variety. One might spend days or weeks in an operation that is virtually in total compliance, and then stumble onto a real "rat's nest" of discrepancies in one isolated area. Conversely, an island of excellence may be found in one foreman's operation, after having enumerated dozens of safety problems in adjacent areas.

In the former case, compliance and excellence seem to be epidemic in the plant, while discrepant conditions were endemic to a single isolated area. Apparently, just the reverse can be stated about the latter situation — an island of excellence existing in a sea of mediocrity and noncompliance. What is prevalent, and therefore a pretty sure indicator of the culture, is what we have dubbed epidemic. What is isolated to a particular area or individual, and stands out because of its rarity, we called endemic.

This obvious truism is belabored here to make an important point. The general prevailing conditions and level of compliance — the epidemic norm, will tend to color the objectivity of the most determined safety auditor. The more accustomed the auditor becomes to either compliance and excellence, or to noncompliance and mediocrity, the less open he or she can become to a subtle trace of the opposite. If sixteen out of eighteen machine tools in a machine shop have the chips cleaned up, the coolant wiped off the floor, the guards all in place, and the operators have their safety glasses on and their hair tied or cut close to the head, the auditor might tend to miss or excuse the exceptions on the other two.

If four out of five shift operators on a catalytic cracker are observed to "fudge" their data entries, the auditor might be inclined to forget the excellence displayed by the fifth operator, who stays on top of the readings.

Yet auditors should not let this happen. The root causes of these two exceptions should be sought — if not by auditors, then certainly by the supervisors, as a result of the safety auditors' reporting of these exceptions. The unsafe conditions noted on a small minority of the operating work stations can cause just as serious an accident or injury as if all eighteen were like that. The individually excellent performance can be raised as a model of "what should be." Only the probability is less — not the severity, when one or two instances of unsafe performance, or unsafe conditions, are allowed to exist. The endemic unsafe condition or work practice is just as unsafe as if it were epidemic. It just won't lead to as many accidents.

It is important for safety auditors to understand the safety culture of an operating area or plant, as we have stressed often before.

However, it is equally important for them to discover those exceptions to that culture, whether they are positive or negative exceptions. It is the deviations, departures, and discrepancies that form the basis of the auditors' recommendations. If there is an endemic problem, it should be the subject of a specific corrective action recommendation. If there is an epidemic problem, it has to be the focal point of a major management system recommendation. On the other side, an endemic positive observation is the signal for an elucidation of why that particular department or foreman should be emulated, and how to do so. Finally, an epidemic positive performance, while a delight to safety auditors, is also the key to correcting the isolated islands of mediocre compliance. They serve as proof that mediocrity need not be the norm.

Little is to be gained from demanding, expecting, or asking for corrective measures against individual (sometimes petty, sometimes significant) findings. True, a specific problem may be corrected, or a particular foreman might be the recipient of accolades, for exceptional safety performance. These kinds of results, however, have only local and temporary effect. The purpose of analyzing physical safety deficiencies (or achievements) is to enhance the compliance of operations, by focusing the responsibility for compliance on the people in charge. They were the ones who have to know that differences exist, that certain elements of their operations are more excellent than others, and, more importantly, why and how. The safety auditors' role in prompting dissemination of this kind of information is inherent in their charter. If they have been chartered to seek the truth, at whatever the price, they can report their findings with impunity, and not worry about the consequences. If, on the other hand, the auditors have been "commissioned" to justify "what is," then the reporting of either type of noncompliance (epidemic or endemic) will be a moral challenge. We are not in a position to advise safety auditors who are caught in such no-win situations, other than to hope that circumstances allow them to respond to their consciences. Situations in which safety professionals are expected to prostitute themselves for enterprises are quite beyond our scope in this work.

Physical observations, as we have stated, are of two kinds. There are the findings in which a requirement is not met, and there are the findings where a requirement is totally compliant. The safety auditors' purposes are to (1) identify both, (2) determine whether a finding is endemic or epidemic, (3) deduce what relevance the finding has in the day-to-day operation, and (4) develop a relevant recommendation regarding the finding. There is nothing quite as injurious to the safety auditors' reputation as a finding that has no merit, or basis in fact, or one that has no suggestion as to how to alleviate it.

Identifying physical plant deficiencies has been dealt with at some length before in this work. Determining whether they are endemic or epidemic has been covered in this section. The relevance of such findings is crucial to the credibility of the auditors, in that the findings, whether positive or negative, must be stated in context of the actual situation observed. In a generally compliant area, a spate of noncompliances must be reported, dealt with, and treated as an oddity to be erased. However, in an area that generally shies away from operational excellence (and safety compliance), the isolated instance of safety awareness and compliance must be treated as an example that needs to be emulated.

A recommendation relevant to the finding, in its particular context, is the last step in making the finding (and its correction) meaningful. The conclusion of a valid safety finding is the formulation of a recommendation that maximizes the value of that finding. If the finding is negative, then (obviously) a positive recommendation countering it is called for. If the finding is positive, a correspondingly positive recommendation is called for, that will extend the benefits beyond the locale in which the finding was made.

We propose that every physical deficiency reported should be accompanied with a totally positive recommendation. Positive or negative, endemic or epidemic, findings are a reflection of "what is" at a given point in time and space. They do not necessarily represent what obtains all day, every day, throughout the operation. So each finding of a physical deficiency, a deviation from the regulatory standards, should generate a positive idea on how to counteract the deficiency, while maintaining the dignity of the operating management and personnel. Only when the problem is determined to be epidemic, should the positive recommendation suggest that management systems be reassessed.

The Real Findings

Physical plant safety findings will be a laundry list of noted deficiencies and compliances. Any item on the checklist might result in an observation worthy of incorporating in the final report. Whether good or bad, these are the product of safety audits—the findings.

So what do we mean by entitling a section "The Real Findings"? The sum total of the findings is a kind of a snapshot of the safety culture in that area at that time. The real findings are the things that derive from an assessment of that body of individual findings, both at the time of the audit, and especially later, during follow-ups and re-audits.

The real findings of safety audits are the general statements that can be deduced from an analysis of the specific findings. They are the conclusions that say "this department has an excellent hazards communication discipline," because during the safety audit, the auditor found all of the MSDSs available on the floor, all the training records current and complete, and sixteen of the eighteen operators interviewed knew the general hazards of the materials that they were working with.

A real finding might be that machine guarding is lax, or considered unimportant, because seven unguarded machines operating in a shop area of 22 pieces of equipment where guarding was called for were found. The real conclusion is that there is a lack of appreciation for the value of guarding, or the existence of production pressures that tempt mere mortals to take chances. Auditors should address this kind of finding when formulating their recommendation for the final audit report.

In this latter example, an auditor can report seven individual cases of machine guarding violations, and make seven corresponding recommendations that the guards be provided. We suggest that this accomplishes little or nothing. Even if the guards are provided, but nothing is done to make it more desirable to use them than not to, they will not be used. If the auditor can, instead, point out the disuse and misuse of machine guards as a cultural or management issue, and recommend management attention to their causes—and *then* provide the guards—whatever drives that disuse or misuse might be removed, and the culture will change.

We have talked in previous chapters about determining root causes. This is where the delving and probing, the series "why" questions, yield reasons that can lead to real solutions to "epidemic" safety problems. By the same token, ferreting out the root causes of safety excellence can provide a number of benefits as well. It will reinforce the management that has fostered that excellence, encouraging them to correct the relatively few instances of noncompliance remaining. It will provide management with road maps for other managers who aspire to improving their safety system, and record. We also bet that it will show that compliance does not cost—it pays.

Our proposition in this section is quite simple. Determine what areas of safety are considered essential, and complied with, and which are considered nuisances or trivial, and therefore ignored or bypassed. Find out *why* certain safety requirements are followed and others are not—what operational pressures cause the operating people to opt in favor of noncompliance, or what cultural forces cause compliance without question. Finally, attack these root causes for noncompliance, and applaud the root causes that make adherence to certain requirements natural.

In our particular industry, the mixing of a number of relatively benign chemical ingredients to make a very energetic propellant is a central and critically important operation. It is a totally remotely controlled operation—no one is exposed to this mixing operation. We would stake our paychecks on the premise that no one would suggest cutting costs by mixing propellants in an attended operation. It is so ingrained in the propellant industry's culture that mixing is a remote operation, that no one would propose otherwise. We cut our teeth on that cultural premise. This is for good reason—if a mixer full of propellant should ignite, the ensuing fire can be both catastrophic and lethal.

Yet such seemingly equally fundamental safety requirements as emergency egress from hazardous locations, control of flammable solvents around propellants, or limiting the quantity of propellants in an area, had not yet, in the recent past, assumed the axiom status of the remote mixing of propellants. Safety professionals in our industry have had to wage these campaigns many times. Why? These matters were not part of the propellant industry's culture. How did we attempt to make it part of the culture? Not by picking at each individual finding, though we made many individual recommendations against specific findings. We accumulated enough instances of the need for corrective action that the responsible management finally understood—*this is a requirement. We'll do it!* We are happy to report that many of these kinds of safety measures are now virtually automatic criteria for new projects, as well as for the upgrading of older facilities.

There is one parenthetical fact that safety auditors, or any safety professionals, for that matter, must remember when a success story happens. It was not the safety auditors that won—it was the entire enterprise. When safety is built into the culture—is the "normal" way of doing business—then everybody, from the stockholders to the janitor, has won. Safety professionals were merely the facilitators of that success. It happens only when their findings are factual, backed by an indisputable requirement, and shown, with data, to need management attention on the broad front.

Management Accountability and Actions

Whatever the condition discovered in the operating area, whatever the real findings might be, it is clearly management that is accountable for the findings, the situation, and for either plaudits or correction. Safety auditors, the safety professionals assigned to the area, the operating people—none of these can effect corrective actions, nor can they claim credit for the good things that auditors might find to report.

Even the excellences that might be found and reported are a reflection on the managers, not the safety engineers who may have influenced them to excel in safety compliance. We have seen instances where the operating crews, who obviously know much more about the safety and hazards of what they do than anyone, have out-analyzed and out-recommended both management and the local safety engineers. Still, it is the managers who must receive the acclaim and acknowledgements when those hazards and requirements are under control. They are the ones who allowed and facilitated the talents and knowledge of their people and advisors to determine what will work best, and implemented it.

We take this rather inflexible position for the simple reason that all of the best advice and information in the world is of no use or value if not accepted and acted upon. By definition, at least in our world, nobody can decide in favor of safety excellence as effectively as the managers of an enterprise. When they build it into the very fabric of the operation, it takes, and sticks. If they only pay it lip service, it is a foregone conclusion that the plant, and its work habits and methods, will reflect lip service.

How safety auditors report their findings is a matter that they must grapple with at the outset, and is heavily dependent on their charter, and the working relationship with their sponsors in management. It will also depend, to a large measure, on how well they size up the managers that they are auditing. As we elaborated on in Part I, if the mentors of the safety auditing efforts are in the ranks of top management, and the auditors feel some confidence in top management's desire to know the truth, however painful, then they probably need not anguish over whether, merely how, to report their conclusions from audits. The auditors can, ideally, feel free to document the conditions observed, draw the "real findings" conclusions that are inferred, and report them in as factual a manner as they can.

If the auditors have been engaged (assigned) for the tacit purpose of condoning "what is," their reporting might take a somewhat different tack, in that they have to be much more factual and diplomatic in justifying their conclusions. We are not alluding in the slightest to professional compromise; we are simply acknowledging a basic fact of life—that how the message is delivered can be just as important, and crucial to credibility, as the message itself. In this latter circumstance, auditors must steer away from personal involvement in the factual presentation of their findings, and lean very heavily upon third party evidence. Not that they are trying to incriminate anyone, they simply have to provide corroboration of their findings, if a reluctant or recalcitrant management is going to accent what they offer, and act upon it.

The safety auditing role is not necessarily a popular one. To the operating people and their supervision, auditors might be viewed as spies and stool pigeons. To the upper levels of management, they may be the unwelcome bearers of bad news that they didn't want to hear. To the safety professionals in the area, they could be seen as the whistle blowers, who are about to advertise the safety engineers' laxity or incompetence. When in an adversarial relationship, these fears and antagonisms are very difficult to avoid. The people itemized above will never be sure of whether the relationship is an antagonistic one, or one of beneficial support, until the final report is issued. The diplomatic, impersonal, and wholly validated factual approach is really the only viable approach to reporting both specific findings and "real findings," the epidemic problems that need attention.

The reporting of positive findings is obviously a less taxing task for auditors. The more kudos that they can deliver, the easier the job of reporting the negative findings. We do not suggest abandoning factuality and diplomacy in this circumstance; we simply observe that a negative finding amongst a host of positive ones is less irksome to auditees. We have known operating managers to actually call our audits "fair" after a report that deservedly contained a factual assessment of both kinds of findings.

So the accountability of management, while an underlying axiom of modern industry, cannot be presumed to obtain when it comes to safety in the workplace. It must be nurtured very carefully, and each and every step that management takes toward safety responsibility and accountability must be documented and reinforced. For the backward managers that we still encounter, it can be a patience-testing process. Hopefully, newly chartered safety auditors will deal more often with progressive, excellence-oriented managers, and not have to cope with issues of management accountability, and willingness to take decisive actions on safety matters. A skillfully structured analogy between (1) the managers' accountability for cost performance, rate and timeliness of production, and the salability (quality) of that production, and (2) the safety of achieving production, is an invaluable tool that safety auditors can use to gradually integrate safety accountability into the culture, where it is not yet understood.

The old saw about actions speaking louder than words is certainly a hallowed truth among safety professionals. Whether wearing the auditor's hat, or that of the on-the-floor safety engineer, or the plant safety manager, these kindred spirits will probably agree on this point sooner than any other. Show us an operating manager who does, not just says, and we will show you a safe operation.

It is the actions that follow the findings that are equally, if not more, important than the findings themselves. Operating managers that take prompt action to clean up their safety problems will make themselves known in the follow-up audit. They will be cleaned up! The difference will be quite apparent. In the happy case where a manager's audit reported little but positive findings, it is less his or her response and action that is significant, than that of the manager's director or vice-president—his or her superior. The latter should be the ones who have picked up on this display of safety integration, and taken action to have it emulated throughout their organization. Upper level managers who fail to applaud and understand isolated cases of excellence in their lower level managers, are indeed safety problems themselves. They have a vastly greater sphere of influence than their subordinates, and can, if they choose, be the catalyst for spreading that excellence much farther than the subordinates can.

Management actions in response to "real audit findings" are the bellweather of the kind of organization that safety auditors are operating in. How they couch their specific findings, real findings, and recommendations will have a measurable bearing on what actions are indeed taken. These findings and recommendations must always be phrased in terms that operating management can accept as blueprints for improvement, not as accusations of present incompetence.

Chapter 16

Identifying Work Practice Deficiencies

We have reached the third head of our triple-headed dragon in safety auditing. Management systems, physical plant, and work practices constitute the triarchy of safety matters, and most of these boil down to the first—management. However, in the less enlightened managements, work practices will still be viewed as a "thing" that evolves independent of stated management goals and altruistic espousals of "SAFETY FIRST." To many of the safety auditors' mentors and clients, unsafe work practices, and, for that matter, unsafe plant conditions, will be considered to be the sole doing of an ignorant and/or defiant work force.

There is a probability of perhaps less than one in a million that plant workers truly choose to perform their job duties in a death-defying fashion. Our purpose in this chapter is to explore some of the reasons behind the seemingly defiant methods that one might observe, and learn to understand what drives workers to do things the way that they do. Work practices are developed from the environment in which people work. Their reasons for compliance or noncompliance are revealed in the attitudes and behaviors of people, in the ways in which they respond to questions about their work practices, and how they rationalize any deviation from what is spelled out in their operating procedures.

The reasons can vary widely; the consequences can vary even more widely. However, the fact is invariant; unsafe work practices can be just as lethal as faulty equipment, or a management that doesn't care.

What Are They?

"Work practice deficiencies" sounds like a reasonably unambiguous phrase, one that won't engender a whole lot of controversial discussion at an ASSE meeting or a company management cocktail party. It sounds like a polite and euphemistic way of saying "operator screw-ups." Well, we are here to tell you that this is a myth akin to those of the Cyclops, the Sirens, and Zeus' lightning bolts.

Work practice deficiencies are the objective comparison of what the procedures say should be done, and what is done in actual practice on the operating floor. We really wish we had another universally accepted term for this discrepancy, other than "deficiency." Deficiency connotes "badness," and many, if not most, departures from written or ordered procedures, do not emanate from "badness," but rather from bad procedures.

The reasons for procedural departures (work practice deficiencies) are several. In our experience, most of them, at least a plurality, are, indeed, bad procedures. Ivory-tower engineers sit in their plush glass-enclosed offices and decree methods and ways of conducting the operation, based on their education and egos, and often little else. They may be extremely knowledgable in the product or component technical requirements. They may be thoroughly familiar with the chemical, mechanical, or physical operations, and all of the engineering science that governs them. However, they simply do not understand the real world of the operational environment, and they usually do not empathize with the doers they are supposed to be instructing.

So the result is operating directives and instructions that are, to one degree or another, simply unusable on the floor. What they lack in practicality, they make up for in unreasonableness. How often we have seen shop planning that is impeccable in form and format, and virtual gibberish when it came to building the product from it.

Another dominant reason for "work practice deficiencies" is the lack of understanding, by workers, of why a given method, sequence, or procedure is important to their safety, and that of others, and the operation as a whole. Translated, this is an utter lack of adequate safety training. The operator who understands not only which buttons to valves to use under which circumstances, but also has a true gut feel for the "why's," "what's," and "how's," will inevitably be a safer operator.

A third basic reason for procedural deviations is the normal, average person's physical inability to do what is written in the prescribed procedures. Physical dexterity, and clearly human limitations, can often restrict the ability of the average crew to perform in accordance with the written word. As often as not, this departure from the "authorized procedures" will look, to safety auditors, like a "work practice deficiency." The obvious message to safety auditors is that it can be beneficial to delve beneath the written word, to determine the reasonableness of the procedure, in light of the objective of the operation.

This is a difficult task for auditors who have no education concerning the operation, who are not familiar with the technology and hazards that attend it. There is little positive benefit, and much negative result, to safety audits that are not totally plausible on the operating floor. In order to assess the "do-ability" of an authorized procedure, safety auditors have to be able to think themselves through it, as written, and satisfy themselves as to whether they could, or could not, perform what the procedure demands. If they conclude that they could, then the auditors may, in fact, have discovered a "work practice deficiency." If they cannot envision themselves performing the described operation as dictated by procedure, even with excellent and pertinent training, then the deviations from it that they will inevitably observe must be couched in the terms that place the cause where it belongs—the procedure itself.

Finally, and we really consider it a rarity, there will be the circumstance where a procedure is violated out of antagonism or a simple desire to do it wrong. The classic case, not very often encountered, of willful violations of reasonable and unambiguous procedures, will, on occasion, surface. People are people, and there those "bad apples" that may test management, and Safety, to see if there are any teeth behind the bark. Existing disciplinary procedures are, of course, appropriate in this case. We caution safety auditors not to be participants to such deliberations, but rather to leave such decisions to management, once their findings are documented and published.

This "documentation" aspect comes up again, and is central to the maintenance of the auditors' credibility. Their notes and reporting of findings could become critical pieces of evidence in a subsequent hearing, at whatever level. We do not intend here to intimidate prospective, or practicing, auditors, nor to launch into a treatise on the various kinds of inquiries that they may encounter. The purpose of this paragraph, and the one preceding, is merely to reinforce the need for factuality in noting and reporting work practice deficiencies.

"What are they?" Work practice deficiencies are the documented departures from what is decreed in the operating procedures. What they mean requires a much more objective analysis of those procedures.

Behavioral Observations

There have been a number of pioneers in the field of behavioral studies as related to the work environment and safety. Two that we have studied, admired, and attempted to emulate in our work are Dr. Tom Krause and Dr. John Hidley. They are both client-oriented consultants and trainers in the field, and prolific writers and lecturers in safety circles. We would either plagiarize their works, or make exceedingly feeble attempts to summarize it, if this were to be a section on behavioral analysis, change, and modification. We highly recommend reading their works to safety auditors who wish to go past auditing, and attempt to effect changes on their own.

However, we will make an ever-so-brief attempt to illustrate the rationale and methods they have developed. Their central premise is that people do things in a certain way (behave) because the "givens" of the situation (the antecedents), and the expected consequences of that way, make it the choice among options. They do not speculate on attitudes, since attitudes cannot be seen and measured. Behaviors (acts), however, can. The antecedents prompt, even force, us to do something. The expected consequences of what we do leads us to choose *which* something. If they, the consequences, are "immediate, pleasant, and certain," then that alternative is what we will opt to do. If they are long-range, unpleasant, and uncertain, we will opt against that choice.

All of this is to suggest that, if there is little or no positive result seen in following a safety rule or directive, it is not likely to be followed, unless the boss is watching. If the rule has an obvious benefit *now*, if it is certain (in the mind of workers) to save them later grief, or if is just plain fun to do it that way, they will do it that way. As we are sometimes prone to do, let's use a stark example to illustrate. You are alone in a burning plane at 10,000 feet. This is an antecedent. You are a moderately experienced pilot, another antecedent. There is a parachute in the cockpit, neatly rolled and packed, and brief pictorial instructions for use, still another antecedent. You have a choice of two behaviors—try to fly the burning plane down to an open field, or don the 'chute and jump. You've never used a parachute before—a fourth antecedent. If you expect the 'chute to open, the immediate, positive, and certain consequence of jumping to get out of the burning plane will dictate your choice of behavior. However, if you don't expect the 'chute to work, or you are certain you won't be able to figure out how to use it, then the negative and perceived certain consequence of leaving the plane for a free-fall to the ground, will dictate to you that you will try to fly it down.

We are not pilots, so to those readers who are, we apologize if this example is too trite or stark to be real. Reality aside, the choice of behavior is seen to result from the combination of the antecedents —the givens, and the expected consequences of each behavioral option that the antecedents offer. We believe whole-heartedly that this is the basis for work practice behaviors too. The rule or procedure has to not only make sense, but make *more* sense than the alternatives. Sense is defined as the measure of how certain and soon a positive consequence is expected.

A more "earthly" example could be a machinist who refuses to wear safety glasses. As he sees it, he can't see as clearly for his high-tolerance precision work. Sweat gets in his eyes and burns. His behavior probably clearly violates a shop safety rule. What are the antecedents? He doesn't like safety glasses; few others wear them, except when the foreman walks through the shop; the rule is there, and he is subject to disciplinary action if he really ignores the rule indefinitely. What about behaviors? There are, obviously two choices—wear them for protection at all times, or wear them only when the foreman is around. How about perceived consequences? If he wears the glasses, the foreman probably won't notice (negative and uncertain), but he won't holler at him either (positive, certain, and soon). He expects discomfort (negative, certain, and soon). He may or may not believe that it might save his eyes (positive, uncertain, and long-term). His fellow workers will likely call him, or think of him as, "chicken" (negative, certain, and soon).

However, if he chooses to ignore the rule, the expected consequences might be; probable notice by the foreman and a dressing-down (negative, uncertain, and unpredictable), comfort (positive, certain, and soon), acceptance by his colleagues (positive, certain, and soon), and the loss of an eye (negative, uncertain, and anything but soon). Putting all of those alternate potential consequences together, and weighing them, as to understanding and likelihood, each individual gravitates toward the option that provides the maximum combination of positive, quick, and sure consequences. If a person does not go through such a painstaking analysis on his or her own, it is still the most overall pleasant and comfortable anticipated result that governs the person's actions. The person does his or her "analysis" by instinct alone.

This is why we blithely talked earlier about "bad" training and bad procedures. They are bad if they do not convince the doers that the consequences of following the procedures, or using the training seriously, are best. Auditors can, if given the time and freedom, do the field research that will provide data on why certain choices are made. They can extract what the expected consequences of each of the options are, in the worker's minds, and thereby get a handle on reasons for work practice deviations. These can then define procedural errors, areas for improvement, or a lack of sufficient knowledge about consequences of a particular action, and identify where intensified, and plausible, training can bear fruit.

There are many other facets and elements to Krause and Hidley's teachings that we will leave to their expertise. One that cannot be ignored is a fundamental, humanistic trait. We all react positively to positive reinforcement, a theme that we have touched on before. They encourage feedback to the working people as to how frequently the right way is used (never mind the wrong way count). The theory is that, if measured against a standard in a positive way (percent compliance), and fairly observed and fed back to the "compliers," then the level of compliance will naturally increase. We cannot validate this theory independently, but we do intuitively believe it to be true.

Whether the reader judges this section on behavioral observations to be, as advertised, "brief," or not, is up to the reader's own determination. We simply wanted to illustrate this basic concept as a possible extension to the more simple audit concept of "what is" versus "what should be," by adding "why is it?"

Endemic or Epidemic, Part 2

We talked at some length in the last chapter about the endemic and/or epidemic nature of physical plant deficiencies. We discussed the distinction between these two descriptions of such deficiencies, and how to treat them in a safety audit report. We do not intend to belabor or repeat that philosophical treatise in this chapter, but merely to apply those ideas to the observation of work practice deficiencies.

"Epidemic" is our term for a condition that is truly pervasive in a plant or operating area. "Endemic" describes a condition—a deficiency, or hopefully, a positive property—that is nevertheless localized to one, or a relatively few, isolated parts, or individuals, within that operating area.

Our premise is that the very same managerial or environmental factors that promote physical plant excellence or mediocrity, will be reflected in work practice excellence or mediocrity. The management style, methods, and approach to the people will influence performance in both the equipment and practice aspects of the operation. That is to say, managers who have production quotas and schedule performance as their primary goals, will be less likely to pay equal attention to the limits of their equipment and people, than those whose goal is maximizing excellence in all of these, and thus maximizing timely productivity, quality, and safety.

The same basic principles apply here as in the preceding chapter. The islands of excellence and maximum productivity that produce epidemic plant compliance will likely have produced work practices bordering on the impeccable. Also, the areas of mediocrity and crisis in a plant operation will probably exhibit slip-shod training and noncompliant work practices as well. Either or both can be lumped together as the result of the environment created, tolerated, demanded, or forced by management.

Yes, management can create an environment of good or poor

work practices by the way in which they manage. If they train their workers in the correct execution of procedures, reporting of problems, and their moral obligation to excel, good work practices will result. If they tolerate sloppy performance, and do nothing but bemoan it, nothing in turn will improve. If, on the other hand, they demand adherence to the documented procedures, and deal constructively with those who miss the mark, overall compliance will inevitably increase. However, if management forces operating people into crutching obsolete and inadequate equipment to achieve the holy production quotas, then compliance will virtually disappear.

Good work practices that are pervasive in an operating area can virtually always be traced to the environment that the operating management creates. Poor work practices—the obvious opposite—are likewise traceable to the management environment created in the area. The epidemic and endemic labels simply derive from the pervasiveness of compliant and/or noncompliant work practices that the safety auditors observe.

"Good" work practices don't just happen; they do not just evolve like the popular concept of the universe suggests—from an inspirational "big bang" of safety awareness. They are the product of a management that truly espouses excellence in every facet of production. How can an enlightened management develop this aspiration for excellence in their operating people? We suggest that it can come about by positive reinforcement. This concept of rewarding the excellent, while basically ignoring the mediocre, is (in our opinion) a powerful motivator for operational excellence. Krouse and Hidley base their theories on the power of feedback, and propose that the feedback of positive observations is much more motivating than the feedback of negative ones.

Their proposition is that feedback on the frequency of adherence to a cardinal safety principle is far more influential than feedback on the frequency of departure from it. An important, even crucial, part of that proposition is that the operating people "buy-in" to the safety principle, rule, or regulation in question. If they do not believe it to be important to comply with a requirement, they won't. Krause and Hidley suggest that the derivation of the operating rules, the tenets of safe operation, be developed by the operating people themselves. To be sure, safety professionals must have their hands in this development, by way of educating the operating people to the hazards, the legal and other indisputable requirements, and the potential consequences of some given solution to a safety problem.

To this end, Krause and Hidley propose a "critical behaviors inventory," or CBIs, that are derived by the crews themselves, along with the help of the safety professional knowledgable in the operation. This provides the "buy-in" so essential to the success of the concept—it is *their* CBI, not the foreman's or management's. A CBI contains a number of do's and don'ts specific to the operation, and is totally inclusive of what the workers believe to be in *their* best interests. The measure of compliance is then made against their own definition of what it is important to comply with.

People naturally like to see positive measures increase, and the measure of compliances is a more positive and powerful motivational tool than the measure of noncompliances, or the measure of loss-producing mishaps. They go so far as to recommend charting the observed levels of compliance for an operating area, rather than advertising the incident or injury statistics per 100 man-years, or the days since the last lost-time injury. The latter are obviously negative feedbacks, while the former is clearly positive.

Whether these principles will stimulate greater compliance levels is an open question, in our experience. We have never had the opportunity to test them in actual practice. We do, nevertheless, intuitively believe that positive reinforcement is a tool that can greatly influence the safety performance of operating crews and foremen throughout industry. We recommend that safety auditors use it as a tool for engendering excellence in the operations that they will be auditing. The input of the doers is, in our view, crucial to improving safety compliance, as well as the quality of the instructions.

So, we now have an additional tool available to assess the meaning of those work practices observed, particularly those that are deviations from the "written word." In the last section, we derived it and called it behavioral observations. The *why* of work practice "deficiencies" is perhaps the most important element of a work practices audit, because it answers the very fundamental questions of (1) what are the drivers, and antecedents, that control or direct behavior, and (2) what are the perceived consequences of a selected work practice, or behavior?

An epidemic noncompliance condition is a very strong signal that the rewards (positive, certain, and soon consequences) of noncompliance are quite powerful, and a serious management and procedural problem demands immediate attention. An endemic, or localized condition of noncompliance would indicate a local management problem, perhaps peculiar to that phase of the operation, or that foreman. Management attention is thereby focused on a lesser sphere of concern, providing their basic approach to people management is adequate.

Of course, by now the reader will understand that "adequate" simply means a problem-solving, facilitating, and reinforcing concept of getting people to maximize their will and determination to do their best. The "dictator" and the "tyrant" will achieve compliance and contribution for a time. However, the will to excel and exceed expectations will never be kindled by such management tactics.

If we seem to be "preaching to the choir" on this point, we will only remind the reader that we stated at the outset that this work was aimed, not only at safety professionals who would be auditors, but also at operating managers who would extract the maximum productivity from their available resources. In addition, we hasten to add, that requires safe production, not just maximum production.

Work practices that are risky are counterproductive, in that they inevitably result in costly accidents, injuries, and property losses. Work practices that are "intrinsically safe" probably result in little or no production at all. It is those work practices that all agree are the safest way to achieve the objectives, that will ultimately prove to be the "right" ones, and the ones that are adhered to. Also, "all" has to include the doers themselves. Without their input, the procedures that are foisted upon them are subject to the same sarcastic acceptance as any other uninformed direction from the ivory tower.

Operating managers who truly want to extract the last ounce of productivity from their people, will be the ones who consult with them about better ways, assesses with them the ramifications on safety of the operation, as well as quality results, production rates and costs, and uses their input, along with that of the safety professionals, to reach decisions that maximize excellence.

Management Accountability and Actions

Who is accountable for the operational performance of the workers on the line? Who is it that should take the blame for sloppy or noncompliant conformance to procedures, or, for that matter, who can take the bows for procedures that work and are followed? We are sure that *our* answers are already anticipated by those readers who have persevered through this work to this point. It is clear by now that we put the responsibility for whatever work practices may exist, squarely on the shoulders of the management of the operation or enterprise.

It is those folks who would manage that *must* accept the responsibility, and the accountability, for whatever goes on in their operating sphere of influence, that is to say, whatever they *manage*. This refrain will be familiar, harking back to the section in our opening chapter on responsibility, authority, and accountability. We still wonder how anyone can see it differently. How can the accountability lie anywhere else than with those who claim the right and privilege of managing the operation?

Unfortunately, there are those who will defy this seemingly impregnable logic, and suggest that the responsibility and blame (accountability) rest with the discoverer of the issues that come out of a safety audit. That is, to reiterate, safety auditors will inevitably encounter those managers who will attempt to shift the accountability for the deficiencies from themselves to the auditors. This can be fairly insidious, taking the familiar form of having safety professionals responsible for reporting the current status of safety problems, planned corrective actions, and their resolution. This trap is quite familiar to seasoned safety professionals, and one to be avoided if at all possible.

We have been there. We know all too well that a traditional management can construe all safety problems to be the responsibility of the Safety Department, and that it is up to them to make sure that the problems "go away." This kind of management is less interested in whether the hazard goes away, than they are in whether their connection with it does. To safety auditors attempting to function in this management environment, we suggest that they attempt to counter this mentality with the simple stance, "if I am to correct it, then simply follow my 'orders'." Depending on the political realities, such a gauntlet might very well have to be tempered a bit to accommodate those realities. However, the message should be sounded clearly, that accountability rests with the ones who are responsible, and have the authority to "issue orders" for the correction of safety problems, which is unequivocably those who are in authority. If that is not the safety professionals, and it hardly ever will be, then logic says to look to someone else for corrective action plans and status.

"Those in authority" is a euphemistic way of identifying the management people who are accorded the privilege of deciding what is best for the enterprise. This may be the operating management for an audit area, or it may turn out to be the Board of Directors of the corporate enterprise. The safety auditors' mission, at whatever level, is to make sure that their findings, the seriousness of them, and possible corrections reach the ears of those "in authority." It does little good for people without the authority to be aware of, or saddled with the responsibility of correcting, the safety deficiencies noted in an audit.

We hasten to add that it might do *some* good for those without authority to nevertheless be aware of those deficiencies. This is true simply because awareness of hazards makes for safer conduct of an operation, regardless of whether there is an overt effort by management to control or eliminate them. The average line workers who are trained, knowledgeable, and aware of specific hazards will "crutch" the procedures and equipment provided, in their all-out best attempt to produce safely. If that sounds vaguely familiar, like "safe production," it is no (pardon the expression) accident. When you get right down to basics, line workers, more than likely, aspire to safe production, even if their management does not.

They are probably more intelligent than their management gives them credit for. First off, they are probably smart enough to realize that production, and subsequent sales of the product, are what provide the funds from which their paychecks emanate. Second, their own safety, whether overtly or tacitly stated, is one of their very highest priorities. Simple self-preservation instinct is extremely powerful and virtually universal in the animal kingdom. Combining the self-preservation instinct with an elemental view of "value added," the average workers are motivated toward "safe production," whether they have ever heard that phrase before or not.

The management that is receptive to the ideas that stem from such an appreciation of work practices and enlightened self-interest will do nothing but stimulate further constructive ideas. The management that stifles self-expression will only see increased indifference, if not outright defiance, and upward trends in their loss-ratio charts. Safe production is an utterly natural, almost subconscious, objective of most workers. If only this were the visceral and unspoken objective of industrial enterprise as well.

Excellence has never been known to be a liability. Yet, as a cornerstone of industrial operations, it is rare enough to be remarkable when it is seen. To the operating managers who may be reading this effort, we simply ask, "why does this apparent paradox exist?" Why isn't the aspiration for excellence the fundamental commandment of *your* operations? If it is, congratulations! If it is not, think about it!

Operational work practices have an unwavering, unchanging property. They evolve toward the simplest, most expeditious, and easiest way of meeting the production goals, in a manner perceived as "acceptably safe," irrespective of the written procedures. This fundamental truth is as unchangeable as the tides—there is no point in fighting it. The one thing that a work practices audit can achieve is, after identifying the work practices that create hazards, to discover why blatantly unsafe activities are seen as "acceptably safe" by those involved. This is about where we started. What are the perceived consequences of a chosen "behavior"? If that crucial question can be answered, then the doers perceptions of consequences can be attacked head-on, and new perceptions developed, through on-the-job training, education (more than just rote training), and enforcement.

We have not had a whole lot to say about enforcement, or the general subject of discipline, thus far in this book. We feel that this is unquestionably the last resort in a manager's hierarchy of methods for promoting and enhancing safety performance. But we must acknowledge that those occasions will arise where disciplinary action, for failure to follow procedures, will become a necessity. We stick to the premise, though, that this ultimate measure should *follow* all reasonable attempts to train, educate, and otherwise control the unacceptable behavior of a worker. However, if it comes to discipline, then it must be sure, consistent, and fair. While this is 100 percent the province of operating managers, safety auditors, or cognizant safety professionals, will probably have some role in this kind of drama, and need to satisfy themselves that the punishment fits the crime, and is consistently meted out.

Safety auditors want accountability to rest in its proper place — with operating management. Just as important, however, they want their corrective actions to be appropriate, fair, and effective. This is why follow-up audits are just as informative as the original audit, and why a continuing program can be so beneficial. The auditors will be able to trace the evolution of management, with respect to their reaction to safety compliance or work practice issues, and determine whether lessons are indeed learned, or simply passed through to subordinates, who lack the responsibility, and therefore the accountability, for the problems.

Chapter 17

Trend Analysis

The ultimate pay-off for safety auditing has several facets. The most immediate should be the identification and quick correction of individual safety deficiencies. We have dealt with this at length heretofore in this work. The intermediate pay-off is the identification of epidemic problems, and the subsequent dissemination of the issues and corrections throughout the whole organization. This, too, has been discussed at length in earlier chapters, as well as this one. The final pay-off comes from the trend of the record for the organization, over time, and the intelligent use of that data to capitalize on the progress made before.

The immediate correction of safety discrepancies and apparent violations of requirements is obviously a step toward improving safety in the operation. Pinpointing particular areas of epidemic problems allows focused attention, by an enlightened management, on that area or operation that seems to have more than their fair share of safety deficiencies. On the other side of the coin, quick reporting of positive safety details, and the focusing of auditor attention on endemic, or epidemic, excellence in the work force, can readily serve to show management that such things are possible, alive, and well in their own organization. It doesn't have to be wished for—it exists and need only be emulated throughout the organization.

Trend analysis takes an even longer, and very objective, look at the progress in correcting not only specific and area deficiencies, but at the progress in changing the entire culture of the organization. It is a data base input and retrieval system that catalogs the problems when and where they are observed, and tracks their alleviation and disregard, as well as their reoccurrence. The very knowledge that this kind of data trend analysis is in being, can often influence the dispatch with which specific and generic corrective actions occur.

Our purpose in this chapter is to provide dedicated safety auditors with a relatively simple way to document *and* store audit findings, so that they can be retrieved for analysis and trend deduction, both on a local and global scale.

Deficiencies and Accidents

One of the most fundamental questions that both management and safety auditors will inevitably ask themselves, and each other, is, "what correlation is there between the accident rate and the audit findings concerning deficiencies?" Obviously, if it turns out that the department with the most, and most serious, audit discrepancies, has (contrarily) the best accident frequency statistics, any rational person would wonder how such a result could occur, whether the audit findings are valid, or whether the accident data might be "doctored." We wish to emphasize at the outset of this discussion that we do not believe there will be any valid, positive, and sustained correlation between safety problems and accident rates in the short term. There is too much of an element of chance involved in an unsafe behavior, or in the use of unsafe equipment or procedures, as to whether that unsafe situation will or will not result in an accident, and how serious it will be. It is only over the course of years, when culture changes, and safe behaviors become an integral part of it, that growing safeness will be reliably seen to beget decreased accidents and losses.

We approach it as an act of faith that the more unsafe the operation is, the more frequent and serious the accidents will be. For those that cannot take that leap, we have only the solace of belief in uncontrollable fate to substitute. Our premise, not original, is that the fewer hazards there are, and the more controlled the remaining ones are, the safer the day-to-day operations will be, *and* the fewer the accidents that will mar that tranquility.

One of the most difficult quantitative problems that safety auditors, and indeed the entire safety profession, faces in the world of industry, is showing and convincing management that their costs are returned with a tidy profit. It is virtually pointless to opine what might have happened had the Safety Department not been johnny-on-the-spot, and prevented some catastrophic accident. For every conjecture in this vein, management has numerous counterpoints when they ask why Safety did not prevent the accidents that have occurred. The only way for safety professionals to avoid, or extricate themselves from, that no-win trap, is to establish the axiom that safety performance is, like quality and cost, a management function and responsibility. Their findings, as auditors, must simply become another data package that management is accountable for, much like the schedule variance report, the unit cost report, and the reject and rework data report.

Safety professionals could learn a valuable lesson from their Quality Control or Inspection colleagues. The quality professionals have generally succeeded in divorcing themselves from any responsibility for the quality of the product in question. They provide management with data, and recommendations for improving quality. If that has a familiar ring, it is (of course) the theme of this book, as it applies to safety. Safety professionals have to somehow divorce themselves from responsibility for the safety of the operation, and assume a role of reporting data, and recommendations for improving safety. As a profession, we have a ways to go in establishing this role as the one that management accepts.

Back to the question — is there a correlation between safety deficiencies and accident rate? Over the long haul, we propose that there must be. In the short term, this will be difficult to support numerically, because of the virtually random nature of accidents. One way to prove the correlation to a "jury" is to accumulate the findings and accident data over a sufficiently long period of time, to allow random chance to assert itself. Another more positive, and quicker, method is to measure the frequency with which "correct" methods and equipment standards are implemented.

This brings us back to our friends, Krause and Hidley, whom we paraphrased in Chapter 16. One of their most fundamental concepts is that positive observations, reported in a constructive way, and measured against their own standards (as deduced from intelligent study of their work situations), will be a more powerful motivator for safe production than all of the safety rules and manuals one can concoct. Consequently, if audit findings are couched in terms of what should be done, and how often it is, we should expect fairly rapid improvement in the frequency of correct operation. Can we really count on this to happen?

If there is a significant positive trend in "correct" operational performance, should there not be an attendant significant negative trend in the accident frequency? Intuitively, we think "yes, of course!" However, to our mutual disappointment, this happy result might not occur. Why? Simply because the "correct" way of doing the operation might not be the safe and efficient way of doing it. How often we have seen legitimate and approved procedures and instructions that virtually fly in the face of safety. The "correct" methods of doing the job may well be less safe than the way it is, in fact, being done on the operating floor.

This is why a real safety audit begins with a real understanding of the operation and its hazards, observes the operation in real time and under real circumstances, interfaces at length with the actual doers, and deduces (from all of the above) what is the safest way of doing the job, what inhibits doing it that way, and what could be done to allow it to be done in the safest, productive way. Which is our introduction to the idea that compliance with established procedures may very well not be the ultimate measure of an operation's safety posture.

With performance data, accumulated from audits and history, safety auditors are in a position to critique the very fabric of the operating norm. If compliance is high, and accident rates are high, too, the conclusion almost has to be that compliance is not enough. There must be something amiss with the procedures being compiled with. There is the distinct possibility that the established way of operating will have to be radically changed, in order to effect any meaningful and significant change in the accident frequency. This concept is generally anathema to the operating management. Status quo is a fairly holy tenet of most managers.

So, where does this lead safety auditors, when it comes to deducing the steps to be taken to reduce the accident frequency? First, they have to know what they are talking about when they cite problems, and suggest solutions. They have to be credible. Second, they have to be able to distinguish real problem situations and behaviors from regulatory and semantic departures, in order to maintain their credibility. Third, they *must* be able to determine why the accidents occur, in spite of near-total or total compliance with the procedures provided. The last of these is the challenge we offer here.

Relating the hazards, and separating them into categories of seriousness, has been dealt with in detail in the early chapters. We have also touched on the subject of the correctness and true meaningfulness of those authoritative procedures. We have driven home the empirical fact that if they are not user-friendly, and readily followed, they will not be adhered to. A "better" way will be found, and used. This is where the data base is valuable. If the data shows that the operating people follow their own procedures, rather than those offered by Engineering, and their accident record is superior to their peers in the organization, then one must conclude that the operating people know safety better than their peers, or the engineers that dictate their procedures.

We will illustrate such a data base later in this chapter. Suffice it to say, in most industrial environments, safety auditors would be figuratively naked without it, if they were to extend their observations beyond the immediate audit in progress, the specific findings of a specific operating area at a specific time. As we have already suggested, the power of safety auditing can be much greater than each localized audit, if the auditors can learn to generalize, document, and (just as important) determine progress, or its lack. True safety consciousness enters the culture when corrective actions implemented by management go beyond telling a machine operator to put the guard back onto the machine, and into the realm where management searches for all the unguarded machines in the shop, and attempts to find out why unguarded machines seem to be preferred by their people. In addition, and just as important, when they applaud and encourage correct, safe production, and enable it.

Record Keeping and Retrieval

The maintenance of impeccable and clear records, and the capability of in-putting/retrieving particular records quickly, are invaluable tools in an effective safety auditing program. While we do not propose that every scrap of note paper be preserved, or every conversation taped, we do recommend allowing a half hour or so at the end of a day, to organize and clarify all of the day's findings. Having these in an orderly, systematic daily record, or file memo, greatly eases the inevitable chore of composing the final report. It also serves as a tidy permanent record in the audit file, composed in real time, dated and initialed, and ready for inspection by the most defensive operating manager.

So, at this late stage in this work, we will discuss the matter of what kinds of records should be kept on the audit findings, how they should be filed or stored, and how we can retrieve specific data we want, to the exclusion of the volumes of data we don't want, at a given point in time, or in response to a given question? There is no one-sentence or one-paragraph answer to that question, so bear with us as we try to develop it.

First, we recommend identifying a method of codifying the several, or dozens, of operating areas that will be audited. For large organizations, with multiple plants and locations, we recommend codification at two levels. For an organization operating at a single plant site, one level may be adequate. Supposing the organization consists of fifteen plants spread over eight states, and each plant has from four to twenty operating departments, or areas. We would identify the plants with a two-digit code—01 through 15, and the subordinate operating departments with a second two-digit code—01 through 20, or higher. Thus, a warehousing operation in the Louisville plant might be coded 0411, meaning the fourth plant on the company's roster, and the eleventh department listed at the Louisville plant. This codification obviously makes electronic data entry and retrieval more easily accomplished.

The next piece of data that we recommend recording is the date and time of the finding. This is not to be nailed down to the split second—that serves no purpose—but in a permanent record, shows when the audit was conducted, and approximately what time of which day the observation was made.

The third piece of data, rather essential to future analyses of trends, is a code that categorizes the finding to a two or three word descriptor. This is a seemingly difficult task, unless auditors have in hand a list of categories and codes from which to select. The category codes should be made up to fit the kinds of operations, hazards, and essential safety programs that the organization is dealing with. Machine guarding is an obvious category in almost any industrial operation, but especially in a machine shop. Material handling is similarly a nearly universal aspect in industry, but especially relevant to a warehousing operation. Categorization of findings is quite open-ended, and should be made up by safety auditors to fit the operations they audit. Under material handling, auditors may wish to distinguish between crane handling operations and forklift, or conveyor type operations. The key point is that the list of categories be representative and inclusive of the auditors' scope of interest.

There will be HazCom, confined space, lock-out, electrical code, and fire protection categories in virtually any enterprise audited. There will be personal protective equipment usage, management involvement, procedure compliance, procedural adequacy, management policy, and system categories throughout industry. However, there will also be a host of categories peculiar to the organization's businesses, also. If high-pressure equipment and processes are involved in an organization's operations, this is a relevant category, whereas it will not be relevant in many other companies. X-ray operations, truck fleet operations, forklift operations, toxic materials processing, forging and press operations, smelting, welding and brazing, railroad and airline safety, and numerous others are examples of industry-specific categories of safety findings. We recognize that each of these can, and should, when the situation warrants it, be broken down into many subcategories. This will be a function of how central a given category is to the operations of the enterprise, and how many *important* nuances there are within the organization, and within our categories. We offer Table 17.1 as a very limited list of categories from which safety auditors might shop, to come up with an initial list of those that apply to their scope of auditing.

The fundamental concept that needs to be understood, in developing a list of safety categories, is that each one describe a quality, or a problem, that will likely be encountered fairly frequently, but is also a matter that good safety sense says needs to be encouraged and expanded, or reduced and eliminated. Clearly, the categories can be both positive and negative. An important safety concern that is controlled through correct and precise operational behaviors, thorough and regular maintenance of equipment, or consistent management support and direction, describes a category where all of the positive findings one can make, will simply reinforce that performance. This leads to the next data entry item, which is a simple designation that the finding is positive (+), negative (−), or, on rare occasions, neutral (0).

Table 17.1. Audit Findings Categories.

Access/egress	Materials labeling
Accident reporting	Materials storage
As-built facility drawings	OSHA records
Confined space	Physical demands
Crane PM	Platform guarding
Electrical code	PM checklist
Emergency planning	PM performance
Equipment grounding	Pneumatic systems
Ergonomic adequacy	PPE adequacy
Exposure data (vapors, dusts)	PPE usage
Fire extinguishers	Pre-operational checks
Fire protection	Pressure systems
Flame control	Procedure adequacy
Flammable handling	Procedure compliance
Flammable storage	Radiation exposure
Hand tools	Safety meetings
Hazards analysis	Safety warnings
Hazcom policy	Shift turnover
Hazcom training	Sound levels
Hazcom understanding	Spill control
Heating/cooling	Storage area control
Hot/cold exposures	Tooling drawings
Housekeeping	Training content
Instrument calibration	Training effectiveness
Laboratory issues	Vehicle issues
Lighting	Ventilation
Lightning protection	Vibration exposures
Lockout	Visitor control
Machine guarding	Warning signs/markers
Maintenance control	Waste handling/disposal
Management communication	Water supply
Management involvement	Work practices (general)
Management policies	Worker perceptions
Materials handling	Workstation adequacy

Obviously, positive findings are the comprehensive policies, open channels of communication with management, facilitating management, safe work practices, well conceived and maintained operating procedures and equipment, compliance to all regulatory and procedural requirements, and all the other things that warm the safety auditor's heart. Negative findings are the noncompliances, whether at management or operating levels, the noninvolvement or nonsupport of management, the deterioration of, or unsafe use of equipment, the absence or nonenforcement of mandatory regulations and standards, and all the other nasty things that auditors find to be contrary to the "what should be's" that they are auditing against. Neutral findings are relatively rare, but are allowed for in the data base, to accommodate those findings that are of interest, but neither to be lauded, nor needing correction. These could, for example, describe a new and different, but compliant and effective way of performing a preventive maintenance inspection, or a qualification test on a new reactor installation, or a new slant and approach to Hazards Communication training.

The final data entry we recommend is the name, initials, or identifying code number of the safety auditor who made the observation and reported the finding in the first place. The purpose here is not to indict the trouble-finder, nor to call him or her on the carpet years later. One way to preserve the identity of the auditor is, of course, to incorporate the daily notes and final audit report into the data base. If this option is chosen, then the data retrieval system doesn't really need the identification entry.

In our view, the audit report, or the succession of periodic audit reports, are the substance of the audit findings. The problem with audit reports only, is that they are usually more voluminous than is convenient, or even reasonable, for extracting historical data on a particular operation or department, or for following trends on any of these over time. The whole purpose of a data base retrieval system is to facilitate the analysis of single, or occasionally double, parameter safety issues. We offer Table 17.2 as an example of a data format that allows retrieval of historical audit findings, and allows trends to be established, using time segments, plants and departments, operational and management categories, and positive or negative nature of each finding as data retrieval parameters.

We do not profess to be expert in the field of data base systems, nor the formulation of entry and extraction programs and methods. Table 17.2 is not intended, therefore, to instruct computer-literate safety auditors on how to set up data bases. It is simply intended to show one of many ways to input/retrieve audit findings data, so that it can be assimilated and analyzed.

We are of a generation that is struggling to integrate electronic processing into our professional efforts. We recognize fully that the

Table 17.2. Audit Findings Data Base.

PLANT DEPT	DATE	TIME	DESCRIPTION	+−	AUDITOR
0411	021989	1435	Electrical code	−	DK
0411	021989	1510	Lockout	−	DK
0411	021989	1545	Crane PM	+	KW
0411	022089	0815	Procedure compliance	+	KW
0411	022089	0835	Hazcom policy	+	DK
0411	022089	0920	Accident Reporting	−	KW
0411	022089	1000	Flammable storage	+	DK
0411	022089	1010	Procedure compliance	−	KW
0411	022089	1025	Confined space	+	DK
0411	022089	1500	Housekeeping	+	KW
0411	022189	0800	Pre-operational check	+	DK
0411	022189	0950	Flammable handling	−	KW
0411	022189	1000	Machine guarding	−	KW
0411	022189	1010	Management policy	+	DK
0515	022689	0830	Flamm. handling	+	DK
0515	022689	0910	OSHA Log	−	KW
0515	022689	1000	Machine guarding	−	DK
0515	022689	1020	Housekeeping	+	DK
0515	022689	1100	Crane PM	−	KW
0515	022689	1110	Lockout	+	KW
0515	022689	0100	Procedure compliance	+	DK
0515	022689	0120	Procedure compliance	+	KW
0515	022689	0130	Procedure compliance	−	KW
0515	022789	0900	Hazcom training	+	KW
0515	022789	0920	Hazcom policy	+	KW
0515	022789	1000	Hazcom understanding	−	KW
0515	022789	1005	Procedure adequacy	+	DK
0515	022789	1020	Procedure compliance	+	DK
0515	022789	1035	Hazards analysis	+	DK
0515	022789	1100	Maintenance control	−	DK
0515	022789	1110	Housekeeping	+	DK
0515	022789	1300	Accident reporting	+	KW
0515	022889	0900	Ventilation	−	KW
0515	022889	1010	Procedure compliance	+	DK
0515	022889	1020	Flame control	+	KW
0515	022889	1100	Machine guarding	+	DK
0515	022889	1310	Machine guarding	−	KW
0515	022889	1420	Procedure compliance	+	DK
0515	022889	1430	Crane PM	+	DK
0515	022889	1440	Maintenance control	+	KW
0515	022889	1450	Flammable handling	+	DK
0515	022889	1455	Procedure compliance	+	KW
0515	022889	1505	Management policy	−	DK
0515	022889	1525	Shift turnover	+	DK
0515	022889	1530	Pre-operational check	+	DK

younger professionals who have cut their teeth on computers and electronic data processing will find the kind of programming we are discussing to be so elementary as to be possibly obsolete. Having explained the data to be recorded, and the potential retrieval parameters that will be of interest in analytical efforts, we will leave the programming, or adaptation of programs, to those who are quite knowledgable about such matters.

Spotting Trends

Trends are tricky! An upward swing in the accident rate is almost sure to be perceived by management as a signal that something has gone wrong. A downward "trend," perhaps over a few weeks or months, causes smiles all around the executive conference room. Are either of these conclusions and reactions justified? More often than not, they are not justified.

An accident is the random result of a repeated unsafe behavior or of an unsafe condition that creates a spectrum of consequences, ranging from nothing, to a near-miss, to the accident itself. How often does a mechanic have to use a faulty tool before an injury occurs? How often does a worker have to smoke around a flammable solvent before a fire is started? The consequences of unsafe actions and conditions are a probabilistic occurrence. For an accident trend to be validated, a very long-term assessment and statistical analysis is required. A period of one, two, or more months with below-average accident frequencies is no valid indicator of a true trend.

Peaks and valleys in the monthly numbers are an indication of a stable and, therefore, unchanging situation. There is no real trend at all, unless plotted or tabulated over much longer periods of time. One way to level the monthly peaks and valleys is to determine a running average. In this method, safety analysts use the frequency rate for the preceding six months as their measure of the accident rate. Then, each ensuing month, they drop off the data for the first of those six months, and add the latest month. If the running average shows a real trend, they can surmise that it might be real.

However, we suggest that plotting and tabulating accident rates as a means of spotting trends is both backwards and counterproductive. A real trend in the safety condition of an operation is more surely, and easily, spotted by measuring and charting the safe behaviors and conditions that exist every day in the operation. Since *safe* conditions do not lead to accidents, while *unsafe* conditions may, at a nebulous probability that on one really knows, tracking unsafeness does not provide the confidence that we all want that a major accident is not imminent. Because the unsafe situation, more often than not, leads to no negative consequence, its existence is not a reliable predictor of negative consequences.

However, because verifiably safe situations do not lead to negative consequences (by definition), and in fact predict such, they are a very reliable indicator of no negative consequences. Our proposition is that the best way to spot safety trends, *and have them acted upon by both management and workers*, is to tell them how often they are following safe procedures and using safe equipment. The more often they do, the safer that they, and the operation, are. Never mind the random fluctuations in accidents—chance occurrences. The upward trend in safeness will eventually, and we suggest, more quickly, be reflected in a corresponding downward trend in accidents.

How do we prove this? We don't. Experts in the behavioral sciences and industrial psychology, whom we have mentioned before, have treated this subject in far greater detail, and with more concrete evidence of the validity of our thesis, than we would presume to. Dan Petersen, Peters and Waterman, and Krause and Hidley are some that we have discussed earlier, and whom we greatly admire and try to emulate. If the reader wishes to learn and understand the intricacies of behavioral modification and worker and management motivation, we defer to those researchers in the field.

We have adopted the philosophy that the people must likely to work safely are the ones who obtain satisfaction and reward from doing so, *and* receive deserved recognition when doing so. When they can expect recognition for doing it right, as well as appropriate discipline or correction for doing it wrong, they will most likely do it right. If they are ignored for doing it right, and castigated for short cuts that backfired, and caused an accident, but were otherwise tacitly accepted when no accident resulted, the temptation will always be there to go with the shortcut.

Trends are tricky! We caution safety auditors to be extremely wary of reporting trends when measured by negative results. If and when the operating management understands and accepts that compliant systems, programs, behaviors, and equipment are the measure of their "safeness," not the vagaries of chance, they too will be more interested in their "safeness quotient." Until they do, auditors will undoubtedly be plied with questions and theories on the accident rate, and why hasn't the audit program reduced it.

Analytical Approaches

We would like to explain a term that will be used repeatedly in this section — indentures, or indenturing. Indentures, the noun, is the computerese term that we, at least, have come to understand as the successive selection of lower and lower levels of an organized data base. Indenturing, the verb, is the act or process of organizing that data base into successive levels of increasing detail. To our computer-literate readers, we apologize for this elementary explanation. To our colleagues who, like us, need to be led by the hand into the world of electronic data gathering, extraction, and analysis, please pay attention, but be sure to engage a programming specialist that can transform these ideas and principles into a useful data base system.

The concept of indenture is fairly simple, once understood, but a bit difficult to explain in words. It allows one to retrieve from a stored body of data (audit findings) all of those findings that relate to one of the recorded parameters. The initial retrieval parameter might be a selected span of time, a particular department or plant, or a particular generic kind of finding. Indenture allows the program to sort beyond that first parameter to select a second parameter for extraction from all of the findings that meet the criterion of the first parameter. If the first parameter, for example, was electrical code compliance, and the second was Department 0411, then programming the system to the second indenture would produce a retrieval of only electrical compliance findings in Department 0411.

This indenture process can be built into the data base system to any degree that safety auditors demand. It can be used to isolate very specific kinds of safety issues in a positive or negative vein, and in a particular operating area, for a specified period of time. For every parameter that auditors wish to add to those already specified, a new indenture level is required.

The purpose of a data base, and the ability to extract data to one or more parameters, is to provide quick and factual answers to questions that management will inevitably pose, once a safety auditing program is in motion, and a number of audit reports have made the rounds in management circles. Believe us, there is nothing to compare with an audit report to spark management's interest in their own posture vis-á-vis "safety."

The data that we have accumulated in a data base, such as the simple one we described in the previous section, will not in and of itself provide analytical deductions, unless it is extracted and studied in a variety of ways. The analytical studies may be directed toward particular questions that are asked of the auditors. One department manager might want to know what kinds of procedural violations are most prevalent in his or her area, and what the most common root causes are. A vice president might want to know which departments, or plants, is most consistent in procedural compliance. A plant manager might want to know which department in the plant has the best housekeeping and maintenance performance. With the simple model we suggested in Table 17.2, this kind of data is obviously not clear from that tabulation. Imagine the length of a printout of a year's worth of auditing.

With the data packed away in chronological order, and from perhaps dozens of four-digit plant/department codes, the first retrieval approach would be to draw out all of the findings for the department manager's request. However, since the manager asked for procedural violations (negative), we would want to sort through his or her total list, with two more parameters. First, we would pick the "Description" code "Procedure compliance" and the "+, −" code "−". If our data base program is capable of sorting to three indentures, we will get a printout of his or her four-digit plant/department for negative procedure compliances.

Depending now on the length of that list, visual inspection may be all the further analysis needed to provide that manager with the entire breakdown of procedural violations, the three or six most common, etc. On the other hand, a lengthy list may have to be further sorted, and this can be done with further indenturing. We must, in order to remain instructional and general, leave the selection of subcategories and degree of indenture to the discretion of auditors and/or their programmers.

The plant manager's question would be answered by extracting, as the first indenture, all findings for each department in the plant. The second indenture, for the first part of the question, would be the description code "housekeeping," and the second indenture for the second part of his or her question would be the description code "maintenance." This will give two printouts for each department. Again, a visual analysis of the $+/-$ ratio may suffice to answer both parts of the manager's compound question. The higher that ratio, obviously, the more consistently compliant the department is—which was the manager's original question.

The vice president's question as to which plant had the best compliance performance would require a data extraction for each plant, sorting each plant's total findings to the description code "procedure compliance" and "+".

We hope that, by now, the concept and basic mechanics are fairly clear to novice data base-conversant auditors, and probably hopelessly juvenile to the computer-literate young professional. The truly competent programmer/data systems technician may wish to record twice as much detail and sort to five or six or more indentures. The more timid or conservative may be satisfied with the rather simple model that we have outlined in the previous section. In any case, we offer one piece of advice to all. The system should be capable of growth and flexibility in the way data is retrieved. The first cut at a list of categories will inevitably grow as audits are added and followed up. New hazards and regulations will describe new generic categories. New products and processes will do the same.

How big a storage capability should be planned? How many audits are planned, and for how long? How many plants and departments are to be audited, for how long, and by how many auditors? We will unabashedly evade this question, and leave it to the professionals who are embarking on the program to design or adapt their data base to their overall auditing plan. However, do give it the capacity and flexibility to grow and undergo modification and sophistication considerably beyond what is envisioned at the beginning.

We offer Tables 17.3 through 17.6 as examples of how the hypothetical questions of the preceding paragraphs might look, if the successive indentures suggested were actually printed out. We will use Table 17.2 as our hypothetical data base, and proceed through successive selections of the data of interest.

Table 17.3. Dept 0515 Procedure Compliances.

PLANT DEPT	DATE	TIME	DESCRIPTION	+—	AUDITOR
0515	022689	0100	Procedure compliance	+	DK
0515	022689	0120	Procedure compliance	+	KW
0515	022689	0130	Procedure compliance	−	KW
0515	022789	1020	Procedure compliance	+	DK
0515	022889	1010	Procedure compliance	+	DK
0515	022889	1420	Procedure compliance	+	DK
0515	022889	1455	Procedure compliance	+	KW

To answer the department manager's question—what are the most prevalent procedure violations in his or her operation?—we extract from plant/department code 0515 all "procedure compliance" codes under description. This gives Table 17.3, which is far easier to comprehend and evaluate than Table 17.2. Now, since the operating manager asked about violations—negative compliances—our third indenture will be "−", which gives us Table 17.4, the procedure violations in Department 0515. The actual result in Table 17.4, a single violation in that three-day period, obviously does not answer the question of what is most prevalent. If one imagines a data base ten or a hundred times the size of Table 17.2, and the subsequent extraction of ten to a hundred times the number of procedure violation observations, then perhaps the value of a data base input and indentured retrieval system can be appreciated.

If there were only ten or thirty instances of procedural violations for the department, it is probable that this level of indenture is adequate. The most prevalent, and the next five most prevalent, can be quickly determined by inspection. At some point, it may become advisable to further indenture procedure violations to more specific kinds, such as doing a step wrong, doing it out of sequence, skipping a step, "graphiting" results, etc. Here again, we leave those decisions to the safety professionals and their data base experts to decide what the situation warrants.

Table 17.4. Dept 0515 Procedure Violations.

PLANT DEPT	DATE	TIME	DESCRIPTION	+—	AUDITOR
0515	022689	0130	Procedure compliance	−	KW

Table 17.5. Dept 0411 Housekeeping and Maintenance.

PLANT DEPT	DATE	TIME	DESCRIPTION	+−	AUDITOR
0411	021989	1545	Crane PM	+	KW
0411	022089	1500	Housekeeping	+	KW

The plant manager's question as to which of his or her departments has the best housekeeping and maintenance record can be similarly answered, as shown in Tables 17.5 and 17.6. Again, the reader should imagine a data base far larger than our minuscule example, where Department 0411 has dozens of findings, rather than just two. The conclusion from Table 17.5 is obviously that the department is perfect in housekeeping and maintenance. The ratio of compliances to noncompliances (+/−) is infinity (2/0). For Department 0515, however, our limited sample tells us that they have a ratio of compliance to noncompliance, or positive/negative findings, of 2. With a subsequential data base, over numerous departments, and over a period of several successive audits and re-audits, a norm will evolve that represents the organizational standard, or average.

Then these ratios become meaningful. Departments may even vie with one another to increase their ratio relative to the others. The point of this discourse is, of course, to illuminate the method by which such data can be analyzed to suit the needs and requests of those who are interested.

Raising the Red Flag

Red flags are historically a warning—a signal that danger is near. Hurricane warnings are announced with red flags—it is too danger-

Table 17.6. Dept 0515 Housekeeping and Maintenance.

PLANT DEPT	DATE	TIME	DESCRIPTION	+−	AUDITOR
0515	022689	1020	Housekeeping	+	DK
0515	022689	1100	Crane PM	−	KW
0515	022789	1100	Maintenance control	−	DK
0515	022789	1110	Housekeeping	+	DK
0515	022889	1430	Crane PM	+	DK
0515	022889	1440	Maintenance control	+	KW

ous to venture out of the harbor. Highway construction hazards are signaled with red flags, red warning signs, or traffic control people decked out in red vests and helmets. Red is simply a historically recognized color for "danger."

Raising a red flag, in our context of safety auditing, means simply warning management that there are problems out there that can lead to serious consequences. There is danger afoot in the enterprise.

It may seem anticlimatic to bring this subject up for review at this late stage of the work. We have labored through the management styles and understanding of the operations, the preparation for, and conduct of, the audit, and how to assess and analyze the findings of the audit. Why worry now about whether or not to report danger signals that are perceived from the audits, or deduced from the analysis of audit findings?

We see this as the final advice to new or veteran safety auditors, that the findings must speak for themselves, and the analytically deduced "big picture" must be reported as the product of a professional analysis. Professional ethics requires that safety professionals, whether auditors, floor engineers, or part-time personnel administrators with the "safety hat," inform their management of any safety concerns that arise in the course of the enterprise's operations. There is no "weasel-wording" that will excuse safety professionals from this duty. Neither is there any excuse for failing to report all of the positive observations that the safety audit is bound to reveal.

Raising the red flag means having the courage to report what was seen, regardless of how irksome that may be expected to be by the operational management. It means having the professional guts to "tell it like it is," whether good or bad. It means not falling into the trap of believing that safety auditors are paid only to find problems, nor into the seductive snare of reporting only what the boss wants to hear. Raising the red flag is our colloquial way of reinforcing the duty of safety auditors to "tell it like it is." Professionally speaking, there is really no other option that meets the test of professional integrity.

There is no excuse for whitewashing a truly hazardous condition, nor is there any legitimate excuse for ignoring the positive elements of a safety audit, or finding fault where none is due. Truly professional safety auditors determine their findings against the requirements, report them as "what is," contrasted with the "what should be's," and retreat, if possible, from the ensuing debate (if any) about the impact of the findings. It is not their role to make the solution easy or painless to the operating management. It *is* their role to make the safety deficiencies unequivocable to *all* of management.

To this end, safety auditors must maintain their independence from, and aloofness to, the politics of the operational environment. The Auditors' role is nothing more nor less than one of providing facts and information to the guys that run the show. Safety professionals are never the ones that run the show, so they should never accept the onus of being the ones that block progress, or get in the way of "process improvements" that are really shortcuts against safety. Their demeanor has to simply be that this is what is required for this enterprise, or operation, and this is what exists. The difference has to be dealt with.

Trends can and should modify the severity, or tolerance with which safety deficiencies are treated. A truly abominable situation that is improving deserves more kudos and less criticism than one that shows no progress from one audit to the next. A truly excellent safety stance that never improves can be open to the safety auditors' criticism also. The message for initiate safety auditors is to couch the findings, and assessment of trends, in terms of relative progress from "what has been" to "what is," and always compared with "what should be."

Any safety professional has but a single duty — to provide accurate information, sound, well thought out opinions, and supportable deductions to management. Safety auditors are more constrained to adhere to these principles than most, because they will forever be seen as the outsiders, and therefore (and without question) suspect as to their motives. As we have said before, the Auditors' most challenging job is to establish an atmosphere where they are seen and received as allies and boosters, and not as natural antagonists. This can only be accomplished through genuine objectivity, whether it pertains to regulations and requirements, or more complex and subjective issues of policy and work practices.

Chapter 18

Epilogue

It is fairly common in mysteries, adventure, or historical novels, for the author(s) to include an epilogue. Its purpose is, of course, to tell the reader, having got that far, what happened after the story ended. Which, of course, becomes a continuation of the story itself. We have said about all we have to say on the subject of safety auditing; the "story" has ended. Yet we feel a compulsion to recap the story, maybe embellish it a bit, and leave the reader with some final thoughts.

Safety Auditing—A Management Tool is a work that was born out of a vacuum on the subject. We have been in the safety auditing business for a number of years, and have had little or no guidance from on high, the lofty towers of academia, or from our management. The concepts, procedures, checklists, protocols, opinions, priorities, and overall approaches to safety auditing are those that we have developed over almost a decade of self-induced trial-and-error. There is nothing mystical about the principles that we have espoused, nor is there anything truly new in our philosophies.

Safety auditing is a relatively new phrase that managements across the country and around the world have learned, or are learning, to talk about, even when they may not know for sure what it means. We do not profess that the methodology described in this work is the only viable one, or that our concepts of operational management stereotypes are the only ones that are valid.

Our purpose, in preparing this treatise on safety auditing, is simply to provide a starting point for this relatively new discipline. That evolutionary development will vastly improve the methods, and their effectiveness, is a given for us. We will not be in the least sorry to see and read a follow-on treatment of the subject that takes us to task on virtually every page. That is called progress and improvement.

So let us reiterate the premise from which we started. In outline form, we propose that:

1. There are numerous styles of management, each of which must be dealt with in a customized fashion, if there is to be any influence on *their* safety performance.
2. It is almost essential that the technology, processes, and materials be understood to a point where auditors can both question and contribute to decisions on safety issues.
3. Conducting audits requires on-the-floor observations, as well as paperwork reviews, in order to distinguish the "what should be's" from the "what is's" and the "what can be's."
4. Finally, the findings must be presented to those who can act, in such a way that intelligent, informed decisions can and will be made.

We have no magic formula for dispelling the age-old nemesis of safety professionals, where (1) they bear the burden of safety problems because they identified them, or (2) the correction of those problems is their responsibility, whether they have any authority or budget to do so, or not, or (3) their sincerest recommendations for safety enhancements are ignored, postponed, or rejected. We hark back to one of our original premises—that the mission of *any* safety professional, of whatever title, is to inform management, and to document that communication.

Since they are not responsible operating managers, safety auditors have discharged all the responsibility that they dare assume, when they have documented their information transmittal to management. They should never fall into the trap of being the messenger who is shot for bringing bad news.

We hope that this work is the opening salvo in a learned interchange of ideas on the purpose, value, merits, methods, philosophies, and importance of safety auditing.

Index

Accidents
  correlation between deficiencies and, 9, 290–293
  definition of, 299–300
  manager's investigation of, 15, 141–142
  and maximum credible events, 81
  spotting trends in, 299–300
Accountability, importance of, in safety, 7–11
Acknowledgements, in problem identification, 238–240
Advisory standards, 94, 107–109
Alcohol, employee perception of, in workplace, 145
American Conference of Governmental Industrial Hygienists (ACGIH), 107–108, 109, 126
American Institute of Chemical Engineers, 96
American National Standards Institute (ANSI), 94, 104, 107–108, 194–195, 204–205
American Petroleum Institute (API), 96, 109
American Society of Mechanical Engineers, 96
American Society of Safety Engineers, 96
Announced/unannounced audits, 115–118
Anonymity, in work practice evaluation, 227–228
Asbestos, vi
Audit. *See also* Safety audit
  nonsafety, 242–244
Authority
  establishing, for audit, 51
  importance of, in safety, 7–11

Bailey, Charles W., 143
Behavior, study of, in workplace, 278–281, 283
Bhopal, v
Blake, R., 25

Checklists
  comprehensive, 110
  for equipment and facilities evaluation, 170–197
  for flammable solvents, 198–199
  OSHA-type, 170
  in safety audits, 67, 110–112, 125
Cited reference, and legitimizing requirements, 93–94
Client, safety auditor's relationship with, 32–34
Coaching of management, 139–140
Commando-type manager, 26, 27
  acceptance of work practices evaluation, 223
  dealing with, 28–29
  guidelines for dealing with, 58
  inaction in, 30
  interviewing, 55, 56
Commitment of management, 132–135
Communication
  acknowledgements in problem identification, 238–240
  on checklist, 183, 196

Communication *(continued)*
  employee perception of, in workplace, 145
  and equipment and facilities evaluation, 183, 196
  explaining the problem, 235–238
  final report, 240–241
    and evaluation of work practices, 227–228
    objectivity and detachment in, 241–242
  with management, 43–45, 139
  in real time, 59, 233–235
Company directive, 94–95
Company size and safety, 31–32
Compliance
  in accordance with operating standards, 214–215
  and communication with management, 44–45
  importance of, 2
  kinds of, 213–214
  OSHA measurement of, 213
  reporting, 212–213
  in work practices evaluation, 211–215
Comprehensive checklists, 110
Compressed Gas Association, 104
Confidentiality, in work practice evaluation, 227–228
Confined spaces
  and equipment and facilities evaluation, 183, 196–197
  on safety audit checklist, 183, 196–197
Conformance to operating procedures, 207–231
Consensus standards, 94, 95, 194–195
Contractual standards, 109–110
Corrective measures, manager's, 15, 142
Credibility, of safety auditor, x, 23, 41, 51–52, 251–256
Critical behaviors inventory (CBI), 283

Data retrieval system, for audit findings, 297–299
Dictator-type manager, 25–26, 27
  acceptance of work practices evaluation, 223
  dealing with, 27–28
  guidelines for dealing with, 57
  inaction in, 30
  interviewing, 55
Disciplinary action, for failure to follow procedures, 287–288

Disclosure, positive handling of, 127
Documentation
  checklist, 67, 110–112, 125
  field notes, 123–124, 202–204, 215–223
  of physical deficiencies identification, 261–265
  published report, 125
  in safety audit checklist, 175–176, 183–186
  of safety delegation, 166
  stand-alone notes, 123, 124–125
  of work practice deficiencies, 277–278
Dusts, on safety audit checklist, 182, 195

Economic advantage to safety efforts, 22–23
  communicating to management, 43–44
Education, of management, 46–49
Electrical assessment, and equipment and facilities evaluation, 179, 187
Employee involvement, employee perception of, in workplace, 145
Employee perceptions, 143
  survey questionnaire evaluating, 147
  topical version, 144–147
  working version, 148–51
Employer, safety auditor's relationship with, 32–34
Endemic problem
  and physical deficiencies identification, 265–268
  in work practice deficiencies, 281–284
Environmental Protection Agency, 66, 94
Epidemic problems
  and physical deficiencies identification, 265–268
  in work practice deficiencies, 281–284
Equal Employment Opportunity laws, 100
Equipment and facilities evaluation, 167
  checklists for, 170–197
    generic, 197–201
  citing requirements in, 204–205
  field notes in, 202–204
  first impressions, 167–170
  recommendations in, 205–206
Ergonomic considerations, on Safety Audit Checklist, 183, 196
Escort requirements, 52, 56
Explosives safety, 176, 186
  and equipment and facilities evaluation, 176–177, 186

INDEX / 313

Facilities and equipment, employee perception of, in workplace, 146–147
Feedback, importance of, in safety audit, 60–61, 282–283
Field notes
  destroying, 203, 204
  and documentation, 123–124
  in equipment and facilities evaluation, 202–204
  in work practices evaluation, 215–223
    elements of, 220
    positive communication of negative information, 220
    positive findings in, 221
First impressions, of equipment and facilities, 167–170
Flammable materials handling, and equipment and facilities evaluation, 181, 191–192
Flammable solvents checklist
  disposal, 198–199
  storage, 198
  usage, 198
Fumes, on safety audit checklist, 182, 195
Fussbudget-type manager, 26, 27
  dealing with, 28
  guidelines for dealing with, 58
  inaction in, 30
  interviewing, 55, 56

General duty clause, of OSHA Rules and Regulations, 96–99
Government contracting and licensing industries, 110

Hand tools, and equipment and facilities evaluation, 181, 190–191
Hazard Communications Checklist, 200–201
Hazard identification
  comprehensive and specific audits, 68–69
  distinction from hazard containment/correction, 91–93
  documentation of, 176, 185
  and general duty clause (OSHAct), 96–99
  and identifications, 230–231
  knowing operations, 65–69
  legitimizing requirements, 93–96
  opinions, requirements and, 90–93
  prioritizing, 14–15, 73–89, 136–137
  and requirements, 90–93
  reviewing paperwork, 72
    drawings and layouts, 71
    management systems, 69–70
    operating procedures, rules, and instructions, 70
    policy statements, 69
    repairs and maintenance records, 71
    training records, 70–71
  targeting, 85–89
Hazard level matrix, 82–88
  matrix levels, 84, 86
  probability levels, 84
  severity levels, 84
Herzberg, F., 25
Hidley, John, 278, 280, 282–283, 291, 300
Humanitarian managers, successfully communicating with, 45

Immediate supervision, employee perception of, in workplace, 146
Improvements, resulting from work practices evaluation, 225–229
Inaction, identifying manager's leaning toward, 29–31
Indentures
  definition of, 301
  and retrieval from audit findings data base, 301–305
Industrial homicide, vi
Information, reporting neutral, 1
In-house standards, 94–95, 106–107
Insurance availability/rates, vi
Integrated Management Control System, 39
Intention
  stating, in safety audits, 118–121
  stating, in work practices evaluation, 217
Introductions, in safety audits, 118–121
Involvement of management, 133–135

Journeyman manager, 25, 27, 29
Judgment, in work practices evaluation, 216–217, 218

Krause, Tom, 278, 280, 282–283, 291, 300

Labor unions, vi, 109–110
Line management
  coaching and improving, 59–61
  educating, 46–49
  interviewing operating manager, 49–50
  checklist for, 50–55
  tailoring interview, 55–58

314 / INDEX

Loner-type manager, 25, 26
  dealing with, 29
  guidelines for dealing with, 58
  interviewing, 56
Loss control, vi, ix
Lower Explosive Limit (LEL), 104

Management
  acceptance of work practices evaluation, 223–225
  accountability of
    and physical deficiencies, 271–274
    and work practice deficiencies, 285–288
  activities
    on-the-floor, 14, 138
    measuring proactive, 135–143
  communication with, 43–45, 139–140
  evaluating, 131–132, 247–260
    checklist for, 154–166
    delegation of safety matters, 152–154
    employee perceptions, 143–151
    management safety systems, 151–152
    measuring involvement and commitment, 132–135
    proactive management activities, 135–143
    subjective, 132–133
  expectations, 1–2
    effect on employee's performance, 226–227
    for purposes of auditing, 18–21
  involvement, employee perception of, in workplace, 145–146
  line
    coaching and improving, 59–61
    educating, 46–49
    interviewing operating manager, 49–50
  methods for influencing, 22–23
  performance evaluation, 247–249
    auditor's role as critic, 249–251
    credibility of auditor, 251–256
    recommendations in, 256–260
  responsibility, authority, and accountability in, 7–11
  safety as a tool in, 3–5
  safety auditing as a tool in, 5–7
  safety auditing programs as training tool for, 41–43
  safety education for, 46–49
  successful communication with, 43–45
Management philosophies/practices, 24–25
  client and employer relationships, 32–34
  dealing with various types, 27–29
  interpreting actions and decisions, 29–31
  company sizes, 31–32
  stereotypical kinds of, 25
Management safety systems, 151–152
Manager
  characteristics of successful, 39–40
  manipulator-type, 26
    dealing with, 28
    guidelines for dealing with, 57
    interviewing, 56
  passive, 15–16, 17–18
    and safety audits, 20, 154
  proactive, 11–13, 14–15, 16, 17
    guidelines for dealing with, 58
    interviewing, 55, 56–57
    safety activities of, 14–15, 136–142
    and safety audits, 19–20, 21, 126, 135–137, 141–142, 154, 226
  reactive, 13, 15, 16, 17
    and safety audits, 19–20
Manipulator-type manager, 26
  dealing with, 28
  guidelines for dealing with, 57
  interviewing, 56
Material-handling equipment, and equipment and facilities evaluation, 180–181, 190
Material Safety Data Sheets (MSDSs), 66, 68, 95, 269
Maximum Credible Event (MCE), 63, 77–80, 98, 103, 167, 186, 208, 217, 230, 237, 254
  causes of, 80–82
  credibility of, 78, 81
  likelihood and seriousness of, 82–85
  and robotics, 79–80
  second-level, 81–82
  targeting the right hazards, 85–89
MBWA activities, 138
MCE. *See* Maximum Credible Event, 63
McGregor, D., 25
Mechanical systems, in evaluating equipment/facilities, 188
Motivation, manager's role in safety, 15, 140–141
Mouton, J., 25

INDEX / 315

National Fire Codes, 108
National Fire Protection Association (NFPA), 66, 94, 107, 108, 110, 126, 194, 204, 252, 302
  Life Safety Code 101, 103
National Institute for Occupational Safety and Health (NIOSH), 102, 108
National Safety Council, 96
Negative information, positive communication of, 220
Noise, on safety audit checklist, 182, 195
Noncompliance, communicating with management that fears, 44-45
Notification of audit, 117-118
Nuclear Regulatory Commission (NRC), 94, 110

Objectivity, in safety audit, 33-34
Occupational Safety and Health Administration Act (OSHAct), v, 97, 100
  inspections, 9
  regulations under, 65, 66, 68, 93-94, 98, 100-106, 125, 170, 195, 199, 213, 214, 219, 236, 302
Operating floor, physical assessment of, 177, 186
Operating manager. *See also* Manager
  expectations of, for safety auditing, 18-21
  involvement of, in safety audit, 59-61
  knowledge of requirements, 90-91
  relationship with safety auditor, 1-2, 3
  safety auditor interview with, 49-58
Operating procedure
  conformance to, 208
  fundamental purposes of, 210
  operational environment's effect on, 208-209
  in work practices evaluation, 207-211
Operational environment, effect of, on operating procedure, 208-209
Operational strengths, identifying, 1-2
Operational success, of safety audit program, 36-39
Operational weaknesses, identifying, 2
Operational work practices, 287. *See also* Work practices
Opinion
  effective communication of, 236
  and hazard identification, 90-93

Outstanding work orders, 184-185
Overaggressive manager, interviewing, 55-56

Passive managers, 15-16, 17-18
  and safety audits, 20, 154
PCBs, vi
Permissible Exposure Level (PEL), 214
Personal Protective Equipment (PPE)
  requirements, 52, 56, 68, 105
  and compliance, 214
  and equipment and facilities evaluation, 179, 187
Peters, Thomas J., 116*n*, 300
Petersen, Dan, 18, 131, 143, 300
Physical deficiencies identification
  determining whether endemic or epidemic, 265-268
  documentation in, 261-265
  management accountability and actions, 271-274
  purpose of, 267
  real findings, 269-271
Positive reinforcement
  value of, in influencing management, 23, 25-26, 128-130
  and work practices, 283
Pre-audit paperwork reviews, 69-72
  drawings and layouts, 71, 75
  management systems, 69-70
  operating procedures, rules, and instructions, 70
  policy statements, 69
  and prioritizing problems, 74-76
  repairs and maintenance records, 71
  training records, 70-71
  velocity measurement records, 75
Pressure/vacuum systems, and equipment and facilities evaluation, 180, 188
Priorities, employee perception of, in workplace, 146
Prioritizing
  hazards, 14-15, 73-89, 136-137
  Maximum Credible Event (MCE), 77-80
  causes of, 80-82
  likelihood and seriousness of, 82-85
  targeting the right hazards, 85-89
  potential problems, 73-76
Proactive managers, 11-13, 14-15, 16, 17
  guidelines for dealing with, 58
  interviewing, 55, 56-57

316 / INDEX

Proactive managers *(continued)*
  safety activities of, 14–15
    measuring, 136–142
    and safety audits, 19–20, 21, 126, 135-137, 141–142, 154, 226
Professional Safety (Bailey and Petersen), 143
Publications, as source of legitimizing requirements, 96
Published report, 125. *See also* Safety audit report

RCRA regulations, 66
Reactive managers, 13, 15, 16, 17
  and safety audits, 19–20
Real findings, in physical deficiencies identification, 269–271
Real-time communication, 59, 233–235
Recommendations
  in equipment and facilities evaluation, 205–206
  in management performance evaluation, 256–260
  in work practices evaluation, 229–231
Record keeping/retrieval and trend analysis, 293–299
  coding operating areas, 294
  safety categories, 294–297
Red flags, raising of, 305–307
Registry of Toxic Effects of Chemical Substances (RTECS), 102
Requirements. *See also* Standards
  citing, as a reason for action, 92–93
  citing, in equipment and facilities evaluation, 204–205
  general duty clause, 96–99
  and hazard identification, 90–93
  legitimizing, 93–96
    advisory standards, 94, 107–9
    contractual standards, 109–110
    industry organizations, 96
    in-house standards, 94–95, 106–107
    OSHA regulations, 93–94
    publications, 96
    recognized expert, 95
    and suppliers, 95
Residual energy, and equipment and facilities evaluation, 180, 188
Responsibility
  definition of, 8
  importance of, in safety, 7–11
Robotics, 79–80

Safe production, planning for, 14, 137–138
Safety
  as management tool, 3–5
  measuring management's involvement in, 132–135
  tacked-on versus built-in, 4
Safety abdication, distinction between, and safety delegation, 153
Safety audit/auditing
  checklists in, 67, 110–112, 125, 170–197
  comparison of, with inspections, ix
  comprehensive, 68–69
  correlation of, with success, 5
  definition of, ix, 6–7
  distinguishing, from safety inspection, 66–67
  effective, 113–114
  elements of, x–xi
  and an evaluation of management, 131–166, 247–260
  findings
    categories for, 294–297
    data base, 298
    positive and negative, 121–123
  first impressions in, 167–170
  hazard identification in, 65–72
  identifying MCEs as central to, 82
  management expectations for purposes of, 18–21
  as a management tool, 5–7
  as a management training tool, 41–43
  notification of, 117–118
  operational success of, 36–39
  paperwork reviews for, 69–72
  physical deficiency identification in, 261–274
  planning for, 67, 68–69, 119–120
  prioritizing in, 14–15, 73–89, 136–137
  protocols and practices in announced and unannounced audits, 115–118
    documentation, 123–125
    eliciting solutions, 126–128
    introductions and intentions, 118–121
    positive and negative findings, 121–123
    reinforcement, 128, 130
  and raising of red flag, 305–306
  as a sampling procedure, 6
  specific, 68–69
  and trend analysis, 289–307
  and work practices deficiencies, 275–288
  and work practices evaluation, 207–231
Safety Audit Checklist

building and area, 176–181, 186–192
environmental and ergonomic considerations, 182–183
generic, 175–176, 183–186
work practices, 181–182, 192–195
Safety auditor
characteristics of a successful, 40–41
and compliance, 211–215
credibility of, x, 23, 41, 51–52, 251–256
as critic, 249–251
and education of management, 46–49
expectations of, for safety audit, 19
as facilitator, 128
hazard knowledge of, 65–72
instinct of, 3
interview of, with operating manager, 49–50
knowledge of requirements, 90–99
as member of corporate staff, 31–32
and policy inequities, 11
and positive reinforcement, 128–130
prioritizing skills of, 73–89
relationship of, with operating management, 1–2, 3
and working with management, 13–14, 19–21
and working with safety engineer, 20–21
Safety audit report, 125, 240–241
distribution of, 120–121
objectivity and detachment in, 241–242
recommendations in, 205–206
Safety commitment, measuring management's, 132–135
Safety deficiencies, correlation between, and accidents, 9, 290–293
Safety delegation, 152–154
distinction between safety abdication and, 153
Safety engineers
expectations of in-house, for safety audit, 20–21
problems of, in conducting safety audit, 31–32
Safety inspections, distinguishing from safety audit, 66–67
Safety involvement
measuring management's, 132–135
Safety meetings, employee perception of, in workplace, 146
Safety problems
generic recommendations for, 253–254
generic solutions to, 250
offering solutions to, 23
Society of Automotive Engineers (SAE), 109
Solutions, importance of offering, to management, 23, 126–128
Stand-alone notes, 124–125
Standards. *See also* Requirements
advisory, 94, 107–109
consensus, 94, 95, 194–195
contractual, 109–110
deriving a checklist, 110–112
in-house, 94–95, 106–107
operating, 214–215
OSHA rules and regulations, 100–106
Success, and safety, 5
Suppliers, as a source of legitimizing requirements, 95

"Tail-twister," 249–251
Teamwork, in solving safety problems, 2
Temperature extremes, on Safety Audit Checklist, 182, 195–196
Theory X managers, 25
Theory Y managers, 25
Threshold Limit Values–Biological Exposure Indices (ACGIH), 109
Threshold Limit Value (TLV), 87
Training
documentation of, on Safety Audit Checklist, 176, 185
employee perception of, in workplace, 148
safety auditing as tool in, 41–43
Training records, in pre-audit paperwork review
Transportation, U.S. Department of, 110
Trend analysis, 289
analytical approaches, 301–305
correlation between deficiencies and accidents, 290–293
record keeping and retrieval, 293–299
record keeping/retrieval data retrieval system, 297–299
spotting trends, 299–300
warning management, 305–307
TSCA regulations, 66
Tyrant-type manager. *See* Dictator-type manager

Vapors, on Safety Audit Checklist, 182, 195
Vibrations, on Safety Audit Checklist, 182, 195

Waste management, and equipment and facilities evaluation, 179, 187
Waterman, Robert H., 116*n*, 300
Williams-Steiger Occupational Safety and Health Act. *See* Occupational Safety and Health Act
Workaholic-type manager. *See* Loner-type manager
Worker's Compensation Laws, vi, 100
Work practice deficiencies identification, 275
  behavioral observations, 278–181
  determining whether endemic or epidemic, 281–284
  documentation in, 277–278
  management accountability and actions, 285–288
  reason for deficiencies, 276–278

Work practices
  and equipment and facilities evaluation, 181–182, 192–193
  evaluation of
    and compliance, 211–213
    field notes in, 215–223
    judgment in, 216–217, 218
    management acceptance of, 223–225
    operating procedures, 207–211
    recommendations in, 229–231
    resulting improvements, 225–229
    stating intention in, 217
  operational, 287
  on Safety Audit Checklist, 181–182, 192–195, 221–223
Workstands, and equipment and facilities evaluation, 180, 189–190